BLEVINS F…
INTERNATIO…

CW01618913

The Bl… Guide to

Living in Spain

EDITION 10

www.blevinsfranks.com

Blevins Franks

Blevins Franks has been providing specialist, professional advice to British expatriates for decades. We have gained a deep understanding of their financial planning needs and of local tax regimes over many years, and have earned a reputation as the leading international tax and wealth management advisers to UK nationals living in Europe.

We provide integrated tax, wealth management and estate planning services to private clients in France, Spain, Portugal, Cyprus, Malta and the UK. These services are provided through a network of over 20 well-established offices, where our Partners live and work in the same geographic area as their clients.

Blevins Franks put our clients first and offer a full range of services which enable our clients to...

- Shelter their capital and income from tax; in particular, tax on investment income and gains, inheritance/succession taxes and (in France and Spain) wealth tax.
- Protect and preserve their wealth and pensions.
- Organise their assets in the most effective manner so that they can be passed to their heirs as efficiently as possible.

Our popular Financial Guides to Living in France, Spain, Portugal and Cyprus have become standard reference books, and are available directly from us.

Whether you are buying a second home in Europe, or going to live outside of the UK (working or retiring), Blevins Franks is perfectly placed to advise you.

We need to talk.

Blevins Franks Offices in Europe

FRANCE	Valbonne Montpellier Toulouse Bergerac Niort
SPAIN	Costa Adeje (Tenerife, Canary Islands) Estepona San Pedro de Alcantara Mijas Costa Nerja Benijófar Javea Barcelona Port de Pollença (Mallorca North) Santa Ponsa (Mallorca South)
PORTUGAL	Algarve
CYPRUS	Paphos
MALTA	Mriehel
UK	London

This book has been prepared based on the laws of the UK and Spain as at March 2014. It is only a general guide and, in explaining complex matters in a simple way, cannot be relied upon as a substitute for professional advice. The authors and publishers cannot accept any responsibility for loss occasioned to any person's acting (or refraining from action) as a result of reading this book. You must take detailed professional advice relevant to your particular circumstances before any action is taken.

The publication you are reading is protected by copyright law. This means that the publisher could take you and your employer to court and claim heavy legal damages if you make unauthorised photocopies from these pages.

The Copyright Licensing Agency (CLA) is an organisation which issues licenses to bring photocopying within the law. It has designed licensing services to cover all kinds of special needs in business, education and government.

If you take photocopies from books, magazines and periodicals at work, your employer should be licensed with the CLA. Make sure you are protected by a photocopying license.

The Copyright Licensing Agency Limited
90 Tottenham Court Road, London W1P 0LP, UK
Telephone +44 (0)20 7400 3100 Fax +44 (0)20 7400 3101

Apart from any fair dealing for the purposes of research or private study, or criticism or review, as permitted under the Copyright, Designs and Patents Act, 1988, this publication may only be reproduced, stored or transmitted, in any form or by any means, with the prior permission in writing of the publishers, or in the case of reprographic reproduction in accordance with the terms and licenses issued by the CLA. Enquiries concerning reproduction outside those terms should be sent to the publishers at the following address:

Blevins Franks Financial Management Limited
28 St James's Square, London SW1Y 4JH, UK

British Library Cataloguing in Publication Data.
A CIP record for this book is available from the British Library.
ISBN 978-99909-3-184-6

Production Management	BPC International Limited
Sub-Editing	BPC International Limited
Typesetting	Progress Press Company Limited
Printed and bound in Malta	Progress Press Company Limited

© Blevins Franks 2014

Preface

This book is now in its tenth edition. It owes much to the Spanish tax advisers, Partners and Private Client Managers at Blevins Franks and their practical experiences over many years of advising numerous expatriates who have moved to Spain.

Since tax regimes and rates are frequently modified, some of the regulations and treatments described in this book may change following publication. We also produce briefer online guides which we update regularly and these are available for you to download from our website – www.blevinsfranks.com

This book could only come together due to the input from a lot of people that came together in a team effort to deliver a successful product.

Particular thanks are due to the advisers in the Blevins Franks Tax Advisory Service team, including David Morley, Nicola Goldsmith, Harri Matikainen, Venita Machnicki, Stephen Morris, Melissa Orwell and Francois Mounielou.

Practical experience and insight of key issues for clients living in Spain has been provided by Partners and Private Client Managers in our offices located in Spain including David Bowern, Terry Wayne, John McCann, Christopher McCann, Steven Langford, Gerald Turner, Neil Jenkins, Robert Croizat, Brett Hanson, Gary Cardwell, Mike Lavin, James Johnson, Paul Montague, Peter Worthington, Rafael Belloch Rodrigo, Kevin Cassar and Alexander Grech.

Due thanks also go to all those people that worked on various aspects such as manuscripts, compliance, style, design and production particularly Amanda Borg Cardona, Sylvana Tabone, Matthew Bonavia and Martha Conti.

Our final thanks go to the team at BPC in particular Kevin Galea for all his patience and attention to detail and to the team at Progress Press for their professional attention to the project, with special thanks to Michel Rizzo and Joseph Schembri.

We hope you will find this book useful.

Preface

Contents

CHAPTER 1

The Legal Side of Buying a Property in Spain

1.1 Introduction

Spain is the most popular overseas country for UK residents buying abroad. Some are buying a second home, others an investment, and still more are buying their dream retirement home. In addition to the Brits, there is a diverse selection of other non-Spanish property owners, so it is a well-developed system.

Purchasing property in Spain is just as safe as buying a house in the UK as long as you take the right professional advice; however, you should proceed with caution. The buying process is very different from that in the UK. You should always engage the services of an independent lawyer that has **not** been recommended to you by the selling estate agent. If you are purchasing the property without a mortgage you should also consider having the property valued independently and not rely on the valuation set by the selling estate agent.

Why be so cautious?

Over the past 30 years, thousands of UK and other nationals have purchased property in Spain. Unfortunately this has given rise to many unscrupulous practices by developers, lawyers and estate agents. Such practices have been highly publicised in newspapers and on television. For example, properties have been illegally built without planning permission and now may have to be bulldozed, leaving the current owners without any compensation.

Properties are registered very differently to procedures in the UK. The title deeds in Spain, known as the *escritura,* are very detailed. They describe the property in great detail to the exact square foot, quoting the number of rooms, patios, pools and even barbecue areas. If the property is not correctly registered, this could mean that parts of the property are illegal and have been built without planning consent.

From our experience we have found that a significant number of properties have registration problems. While in the past this may not have been a problem at resale, today buyers are more informed and often require mortgages. Banks are only prepared to lend on correctly registered titles.

1.2 Estate agents

There are many reputable estate agents in Spain who are licensed by the *Agente de la Propiedad Inmobiliaria (API)* or by the *Gestor Intermediario en Promociónes de Edificiones (GIPE).* They can provide an experienced guide to the area in which they operate. The estate agent may be able to recommend architects, surveyors, banks, mortgage lenders and other contacts in the area. However, this is a de-regulated industry in Spain, so almost anyone can broker a property sale in Spain, and you should choose your estate agent with care.

1.3 Lawyers

To purchase a property in Spain you do not have to use a lawyer to represent you at the completion meeting which is held in front of the public notary. The role of the notary is to ensure that you understand what you are signing. They are not there to check that the property you are purchasing is legal and correctly registered. If an estate agent tries to talk you out of using a lawyer or into using their recommended firm, this may be an indication that there are problems with the property.

There are lawyers in the UK and English speaking ones in Spain who are aware of the issues and differences between the UK and Spain, and so can explain the process in terms you understand and guide you expertly through the process. The lawyer in Spain is called an *abogado/a.*

1.4 Valuations

When visiting a property, the valuer carries out a number of measurements, tests and searches in order to establish the value of the property. The valuation report points out the legality of the property in accordance with current legislation. Under Bank of Spain rules, the valuer can only value what is recorded on the *escritura,* so if the property has been extended and the extension has not been properly recorded it cannot be included in the valuation. The valuation will be checked against the *escritura* by your lawyer to ensure that the property is correctly registered.

It should be noted that, while estate agents generally provide an optimistic appraisal of property in order to secure the sale, in the past they have relied on purchasers buying property for cash without having an independent valuation. This has led to some individuals overpaying for properties in the past.

The value of rustic properties will usually be lower than that of urban properties. Spanish valuers will for the most part have a different view from that of foreigners. For many expats, the thought of owning a villa in the countryside with open views and nothing in the vicinity to cause noise and disturbance is part of their dream. The Spanish viewpoint on the other hand is that nobody in their right minds would want to live out in the middle of nowhere with no main roads, shops and amenities. Hence the reason for low land prices in rustic areas.

1.5 Mortgages

If you plan to take out a mortgage to finance the property, either against a UK or the Spanish property, it is better to obtain this at an early stage. There are two options:

1. Against your UK property; the mortgage will usually be in Sterling; *or*
2. Against the Spanish property; here the mortgage will be in Euros.

Interest rates, methods of calculating how much you can borrow, and the costs of borrowing vary between the countries, and also between lenders in each country. Of course, if you borrow in Sterling you run the risk that the Euro may weaken against the pound, and you lose out although, the opposite may happen too.

If you are going to let the property, you are unlikely to obtain tax relief in Spain on a UK mortgage.

The main differences between an English and a Spanish mortgage

- Spanish mortgages are available on a capital and interest (repayment) basis or on an interest-only basis.
- It is common for interest rates to be offered on a 'variable' basis whereby the interest rate is linked to an interest 'base rate' such as the EURIBOR (Euro Inter Bank Offered Rate) or the Spanish *Índice de Referencia de Préstamos Hipotecarios (IRPHCE).* Fixed interest rates may be available for terms of up to 30 years. However, fixed rates are rare in Spain and are normally only reserved for capital and interest (repayment) mortgages.
- Most Spanish mortgages are usually granted for up to 40 years, subject to the mortgage term ending before the eldest applicant's 70th birthday.
- There are often penalties imposed for early repayment of the loan.
- The mortgage will be arranged in Euros and the bank will secure a charge over the Spanish property.
- The maximum loan is generally 80% of the value of the property or purchase price, whichever the lowest. You must therefore make sure that you have adequate funds to pay your deposit on the property.
- Few lenders in Spain work on the UK model of income 'multipliers', and more common is the European 'affordability'

approach to lending. Typically, the calculation will make sure that the total of the Spanish mortgage payment plus UK mortgage/ rent, plus any long-term borrowings should not exceed a set percentage of **net** monthly income.

1.6 The Spanish notary *(notario)*

The notary is a public official who has no real equivalent in the English system. Their duties are similar to that of an English solicitor, but the notary is concerned with making sure the legal requirements have been met in full. The notary has no responsibility to either party, and will not usually advise you on your purchase, but is responsible for checking that the papers are correct and that certain government regulations have been satisfied, including in certain cases the payment of tax.

As a purchaser, you should therefore always ensure that you have an independent lawyer acting on your behalf, and not for the seller. The buyer's solicitor will check the title and take care that there are no arrears of electricity, community fees, taxes and so on (this is very important, as the buyer effectively takes over the seller's unpaid local taxes, including interest and penalties due). The lawyer acting for you should also explain everything to you in English, and therefore it is helpful either to have an English lawyer who specialises in Spanish properties, or a Spanish lawyer whose English is very good. For a new property, your lawyer should check at the Town Hall that the developer has obtained proper planning permission.

The key document in Spain is called the *escritura,* which is the deed of title and must be approved and witnessed by a notary before it can be registered.

It is very important that your title to the property is registered in the public registers in Spain. If it is not, then you could have considerable legal difficulties later.

The main register is known as the *registro de la propiedad.*

It is usually far better for you to have the *escritura* translated from Spanish so that you can understand it in full.

Once the property has been registered, the *escritura* becomes an *escritura pública,* which is a formal publicly registered title deed. Only when you have this piece of paper can you easily prove that you are the true owner of the property.

1.7 The surveyor

Your lawyer or estate agent can recommend a surveyor. Spanish surveyors tend to produce reports that are somewhat shorter than those of an English surveyor and with different content. You may want to spell out in more detail the kind of report you are expecting, if you are using a Spanish surveyor; alternatively there are UK-trained surveyors working in Spain who could be used instead.

The sort of things you might consider your surveyor checking are: the adequacy of the foundations; the utilities (heating, air-conditioning, pool, septic tanks); the electricity supply; the presence of insects; whether there is asbestos in the building; and the condition of the roof.

It is just as important to have a survey done in Spain as in the UK, so you know exactly what you are buying.

1.8 Contracts 'subject to survey and enquiries'

In Spain you do not have a contract which is 'subject to survey'. Therefore you should make sure you commission a survey and are satisfied with its results before you sign the contract.

Similarly you should have undertaken your enquiries beforehand. Your solicitor can guide you through this. You might want to know what the plans are for the town, whether there are new developments, an airport to be built, whether motorways are to be built, and if so, where, and so on.

1.9 Title insurance

This is becoming more and more common, and there are several companies who will insure you against various issues arising regarding the title to the property, such as the lack of planning permission or illegal planning permission, or debts against the property.

You should remember, that while this may be beneficial, it is no substitute for proper legal checks made by a lawyer (and in fact, the insurance company is likely to make similar checks regarding the title and registration of the property). It is also an insurance product, and you should ensure that you read the small print to establish exactly what is covered, and the circumstances under which the policy will pay out.

1.10 Registration

Under Spanish law, debt can attach to a property. In other words, if you are not careful and the notary and your solicitor have not done their job properly, it is possible to buy a property which has debt attached to it. You have in effect taken over the debt of the previous owner. That is why it is vital that your lawyer checks the registration with the public land registry since, if there is no debt registered, you are safe.

1.11 Under-declaration of price in the deed of sale

If you buy a house in Spain, you might be asked to break the law as a routine part of the process, although this has become much less common in recent years. In order for the seller to reduce their capital gains tax, they will ask you to declare that you are buying the house at a certain price that is lower than the real price agreed, but ask you to pay the difference in cash. This is called black money. You will be assured that it is perfectly normal in Spain, but this is not the case.

The Spanish authorities have been clamping down on under declarations in recent years. While this practice may still happen in certain areas, you should resist. Unfortunately, it might mean that you will lose the house of your dreams, but you should be prepared to walk away from it.

The Spanish tax authorities are well aware that this crime is committed, and they are putting more effort into tracking down those involved. There is a computer program used by notaries, property registry offices and the tax authorities to record and review all property sales, and which allows them to compare prices locally, to see if sale prices are realistic.

As a buyer, if you were to pay cash under the table, it means that when you come to sell the property, you will only be able to use as your cost deduction the declared value. You may then end up having to pay tax on the capital gain made by your seller. That could mean an extra tax liability of up to 27% on the under-declared value in Spain when you sell the property.

Payment of black money is now harder than ever, since laws have been introduced that force buyers to declare in the purchase documents how they are paying for the property (ie, bank transfer, mortgage, cheque or cash) so that bank accounts can be checked to verify the price actually paid.

Furthermore, if any estate agent, lawyer, adviser or bank manager in the UK is aware or suspicious that you are paying cash under the table, they must report the transaction to the Serious Organised Crime Agency (SOCA), who will in turn pass the information to the Spanish authorities.

1.12 Payment of the purchase price

Payment is usually made in front of the notary. If you are buying from a non-Spanish resident you are obliged to retain 3% and pay it to the Spanish tax authorities in respect of the seller's capital gains tax liability (the notary will usually do this).

1.13 Important documents and terms

Licence of First Occupation (LFO)

One of the most important documents that you need to be familiar with is the Licence of First Occupation (LFO). This certificate shows that the property fully complies with all current planning and building legislation. It is granted by the Town Hall *(ayuntamiento)* and shows that a property has been built in accordance with the original building licence which was also granted by the same Town Hall.

In order to issue the licence, the property is inspected by chartered surveyors who will certify that the property complies with health, safety, planning and construction laws. The surveyor will also confirm that the property has been fully completed and that there are no outstanding works.

For the utility companies to provide connection to water, electricity, etc, the licence will be required. Some utility companies will proceed with the connection of facilities on receipt of the application of the licence, depending on the region.

There could be many reasons why a property does not have an LFO and in some areas, these licences have been granted by the Town Hall even if the property does not comply with the law.

The LFO is also known as the *cédula de habitabilidad, cédula de ocupación* or *cédula de ocupación (segunda vivienda).* In order to obtain an LFO it is necessary to contact the *departamento de urbanismo* at the Town Hall and ask for a copy, as the originals are not provided. You should ask the seller to provide the LFO.

To request the LFO, the following documents will be necessary:

- *NIE* certificate/passport.
- *Escritura* showing ownership of the property.
- A plan of the development/property.
- An ordnance survey plan of the area.
- The planning sector reference of the development.

These last three are contained in a book at the Town Hall and charges are involved. LFOs have been in existence for about 35 years.

If the LFO was not issued in accordance with the law, your property may be considered illegal. Even if a building licence was issued, the property might not comply with planning laws. Some properties might not comply with the *PGOU* or town urban plan.

IMPUESTO SOBRE BIENES INMUEBLES (IBI)

Another important document is the *IBI* certificate. This is the municipal real estate tax, very similar to the UK council tax and is based on the *Cadastral* register. While this does show that the Town Hall has a record of the property and is providing services for it, the *IBI* in no way substitutes the LFO or establishes the property as complying with the law.

NOTA SIMPLE

The *nota simple* should show the registered owners of the property, present classification of the land, any charges, mortgages, rights of way, taxes or embargoes that may be on-going or have lapsed. This document is basically an extract from the title deeds register.

PLANNING CHARGE

A planning charge may arise where a developer was liable to give extra guarantees to the Town Hall for work to be done in order to obtain the LFO. For example, this could be for improving the infrastructure in the area surrounding the urbanisation. As an owner of a property you could be liable for a proportion of the costs involved in undertaking such work and therefore the charge appears on the *nota simple.*

RIGHT OF WAY

A right of way refers to pipes and utilities passing through a property. The right of way gives access to the developer in order to carry out repairs or improvements to the utilities. Some rustic properties may give right of way for animals to pass through land attached to a property.

MORTGAGES CANCELLED BUT NOT REMOVED

When taking out a mortgage in Spain, a mortgage deed is signed before a notary which is then inscribed in the land registry as a charge on the property. Unless the mortgage is cancelled both economically and at the corresponding land registry, this will still show on the *nota simple.*

In order to remove the charge, signing an *escritura de cancelación* at the notary and registering it at the land registry will be necessary.

REGISTRATION

To declare and register a new build *(declaración de Obra Nueva)* a deed would be necessary to duly register the property. It may also result in the need for both planning permission and an LFO. Some properties have extensions of major dimensions which could be considered significant and in these instances a declaration also needs to be made.

Sometimes the title deed can show that the property is an office even though in reality it is used as a house, is 'in ruins' or 'under construction'. In such cases, it is very important to register the property correctly.

In some cases a property could have existed for many years but was built prior to the existence of the land registry or was never registered and therefore does not exist in their records. It would then be necessary to obtain a *declaración de Obra Antigua.*

CONSOLIDATED BY AGE CERTIFICATE

In Spanish planning law there is a rule which states that a relevant authority has four to six years (depending on the area) from the date of an infraction of the law to notify the individual who caused the infraction of the duty to return the affected area to its original state. In reality, though, the local authority only has to record its intention to take action against the property within the allotted timeframe. The notification may not be issued until they begin actual proceedings.

If, after the relevant period of time has expired, there has been no record of intention or notification by the authority, the breach (eg, an extension) is nullified and no further action can be taken.

If you are purchasing a property that has been extended and these alterations are not registered on the *escritura* you should ask the seller to provide you with a certificate from the Town Hall stating that the property is 'consolidated by age'.

Please note that this rule does not affect certain land classifications or types which can never be built upon legally and are therefore always subject to the possibility of future legal action.

SUBSIDISED DWELLING *(VPO)*

If a property is a *VPO* it means that it is a state subsidised dwelling for very low household incomes. The property is not priced by the free market. Rather, laws stipulate at what price the property is to be sold, which is usually below the market price of a similar property. The *VPO* does have an expiry date after which time the property registry can be updated and the property sold at market value.

JUDICIAL CHARGE

A charge can be the result of a judicial decision and a creditor can place the charge on any one or all of the properties if they so wish. That is, if they wanted to they could place the charge only on one property which is part of a development and the owners of that property would therefore be liable for the full amount. Alternatively, the creditor could spread the amount across all the developments and thus each home owner would have a smaller liability.

The recommended course of action is for the purchaser to ensure that all charges are cleared before proceeding.

RUSTIC LAND

There are two types of rustic properties: common rustic land or specially protected land. Generally speaking, rustic land is set aside for agricultural activities and therefore should not be developed unless the local authority has given express approval and issued the relevant building licences. If the land is 'specially protected', proceed only with extreme caution. In many parts of Spain it is illegal to build on this type of land.

Another aspect to be wary of is where an old property such as a dilapidated farm house or barn is either converted or torn down and rebuilt as a villa. In this case the prospective purchaser should show caution as often approval from the local authority has not been sought and therefore the property is unlikely to be registered correctly. In fact, the property may remain registered as a barn or, worse still, as a ruin.

IN SUMMARY

The purchase of a property is a substantial investment and commitment. When purchasing in a country where you have little knowledge of the legal system, the extra costs of engaging an independent lawyer and valuer will ensure that you are protected and are not left with a property that could be difficult to sell at a future date.

1.14 Who should own the property?

There are several options:

Joint names

This is the usual method for a married couple. Although Spanish succession law forces part of the 50% owned by a deceased parent to pass to the children, this succession law does not normally apply to foreigners, provided they have made a Will that is valid in Spain. However, Spanish succession tax may be payable on the death of a spouse (unlike in the UK, this is not a tax exempt transfer) even if you are not Spanish tax-resident.

Adding children's names

This can save Spanish succession tax but remember that:

- On death, the gift can be treated as a 'gift with reservation of benefit' in the UK for inheritance tax, and so remains in your estate for UK tax purposes, even if it is not yours for Spanish tax purposes.
- You may be liable to pay the UK 'pre-owned assets' tax if you are UK resident and have use of the property.
- If your child divorces, becomes bankrupt, or you fall out, this will have adverse consequences for you.

- The child should be over 18, as it is very difficult to sell the property if it is in a minor child's name.
- The child could insist on a sale of the property in order to receive their share.
- Beware that the gift to your children does not give rise to an immediate liability to gift tax in Spain. Ensure the child is given his/her share of the cost before it is remitted to Spain, and that it genuinely comes from that child's bank account, as giving a share of the property itself may create the very tax charge you are trying to avoid.
- You can reduce these risks by having a usufruct (a life interest) whereby you have the right to use the property rent-free during your (and your spouse's) lifetime, although this could still have implications for UK inheritance tax.

Limited company

This is the most expensive to set up in general. As well as set-up costs (which are higher for a Spanish company than a UK company), there will be annual compliance costs, including tax returns for the company. In addition to any corporation tax due on income or gains derived from the property, personal income tax will be due on extraction of the profits from the company.

A company incorporated in Spain may actually be resident in the UK, if the directors and 'effective management' of the company is in the UK, and therefore UK corporation tax may be due on any income or gain derived from the property.

If you use a tax haven company (eg, situated in a low or no tax jurisdiction) there is an extra 3% tax to pay **every year** on the value of the property.

You may find that you are liable to tax on the 'benefit in kind' provided by the company if you live in or have use of a property owned by a company if you are Spanish resident.

Succession tax

For wealthy individuals owning expensive properties in Spain, an offshore company structure can be recommended to save tax. Professional advice should be sought as such companies themselves give rise to significant tax liabilities. They are generally of no use for the ordinary buyer of an ordinary home.

The only way to reduce this liability for non-residents of Spain is to take out a Spanish mortgage over the property, although this will incur other costs, such as mortgage fees, etc, and interest charges will, of course, need to be taken into account.

1.15 Tax identification number *(NIE)*

There are three types of Spanish tax identification numbers:

1. *CIF* for companies.
2. *NIF* for Spanish nationals.
3. *NIE* for foreign residents or foreigners wishing to buy property in Spain. If you reside in Spain as a foreigner, or if you own a property and are non-resident, you need a *Número de Identificación de Extranjero* or *NIE.* Most significant transactions require you to provide this fiscal identification number. The number identifies you to the Spanish tax authority and is required when you pay your taxes or have any dealings with the *Hacienda.*

Previously, it was possible to buy Spanish property prior to obtaining this number; however, changes in the law now require all foreigners buying property to secure their *NIE* and the accompanying letter before completing their purchase. An *NIE* is also highly important for any expat living in Spain as it is needed when: setting up utilities; applying for a Spanish driving licence or validating an EU one; applying for a business permit; any dealings with the tax authority; inheriting any Spanish assets, and making purchases above €3,000.

To obtain one, you need to present yourself at the nearest police *comisaría* (with a foreigner's department – *Oficina de Extranjeros*) with the following documentation:

- A completed and signed application form and a photocopy of the completed form.
- Your passport (and a photocopy).
- Proof of address in both the UK and Spain (if you do not have a Spanish address, that of a friend, solicitor or estate agent will do).
- Written justification of why you need the *NIE*; this can be issued by an accountant, notary, solicitor, estate agent, bank manager, insurance agent, former employer, etc.

The *NIE* will take two to six weeks to come through.

You can also apply to the Central Police Headquarters in Madrid or to your local Spanish consulate while living outside Spain. The above documents are required, along with two recent passport photos, an SAE addressed to the Central Police Headquarters in Madrid which is big enough to contain all of the required documentation, plus an application form.

An *NIE* is highly important for a Spanish resident so it is important to make sure you keep track of it and it is advisable to make several photocopies.

A non-Spanish citizen needs an *NIE* once they reach the age of 18 (it is age 14 for citizens).

1.16 The community of owners *(Comunidad De Propietarious)*

Most people who buy a property in Spain will be members of a community of owners, which is a sort of tenants'co-operative. Even though the tenant will be the freehold owner of their property, there will usually be common parts of a development (halls, staircases, patios, swimming pools, gardens, etc) owned equally by all tenants as 'community property'. The law is governed in Spain by the Law of Horizontal Ownership.

This law lists the responsibilities of the community of owners, which are:

- The property must be maintained in a good state of repair. The owner of each property can be liable for paying any damages to the community property.
- The owner must take care of any main water pipes, drains, heating pipes and electric wiring which run through the property.
- Community expenses *(cuota de participación)* must be paid on time. If not, the community has a legal right to embargo your property.
- Whenever it is necessary for work to be carried out and access is required into any house or flat, then it should be given by the owner.
- The tenant must abide by the rules of the community.
- The tenant has the right to vote and to speak in any meeting called.

Each member is allocated shares according to the size of their property. Each community has its own set of rules. The number of shares determines not only the size of your contribution to the cost, but also the number of votes.

The members themselves elect a chairman *(presidente)* who is legally responsible for the community and seeing that it obeys Spanish law. The members may appoint an administrator and even a secretary and a treasurer. The chairman has to be an owner, whereas the administrator, secretary and treasury need not be. The law only requires the appointment of a chairman.

The chairman is also responsible for ensuring the smooth running of the community. This includes arranging for the repairs and maintenance of the common parts, drawing up a budget for such purposes and collecting each member's share of that budget. Usually, annual accounts are prepared.

Each community has an Annual General Meeting. You may either attend yourself, or appoint a proxy.

The annual meeting elects a chairman for the coming year, approves the accounts and budget, fixes each member's contribution for the forthcoming year and deals with any other matters. Resolutions are passed by a simple majority of votes provided that at least 50% of the voting power is represented. The meeting can also vote to change the community rules but such votes must be carried unanimously. Anyone not attending is given notice of this change and, if they do not disagree within one month, the change will be effected. If you do not pay your community charges, your property can be charged and ultimately sold at auction to pay the unpaid fees. When a flat or house is purchased, you should be given a copy of the rules *(estatutos)* of the community.

1.17 Power of attorney *(Poderes)*

If it is not possible for you to go to Spain to sign the *escritura* you can give a power of attorney to someone else to do so on your behalf. The power of attorney will be signed in front of a notary either in the UK or in Spain. If it is signed in front of a UK notary it has to be ratified by the Foreign and Commonwealth Office for use overseas.

The type of Spanish power of attorney that you will need depends on what you want to use it for. Your lawyer can discuss your requirements with you and prepare the necessary document. Alternatively, you can deal directly with the Spanish notary who will ultimately need the power of attorney.

1.18 Costs of buying

Notary fees

The notary's fees are set by law on a sliding scale, depending on the declared price, the size of the property and the number of pages and clauses in the *escritura*. Notary fees are subject to VAT *(IVA)* in Spain.

As a general guide, if you wish to avoid detailed calculation, allow up to 0.5%. If you are taking out a Spanish mortgage, notary fees (and VAT) will also be due on the mortgage, depending on the amount.

If you have asked the notary to do any additional work over and above the transfer title to the property, or to provide any advice, there will be additional charges.

ITP

On buying resale property or a property from a private individual, you will also have to pay a transfer tax known as *Impuesto sobre Transmisiones Patrimoniales (ITP)* of, generally, 7% of the value declared in the contract (although some regions have increased this to 8%), similar to UK stamp duty land tax. The rate varies in different parts of Spain.

IVA

Spanish VAT *(IVA – Impuesto sobre el Valor Añadido)* at 10% is payable on new build residential properties (eg, acquired from a developer), or resold, unregistered properties (ie, an off-plan property acquired from another individual prior to being registered). Commercial properties and car parking spaces are subject to VAT at 21%.

Plots of land are subject to VAT at 21%, but if you commence building a residential property on the land, you can reclaim some of the VAT paid, bringing the VAT rate down to 10%.

There is no VAT in the Canary Islands; instead, a sales tax (the *IGIC*) of around 5% is levied.

Stamp duty

Stamp duty *(Impuesto sobre Actos Juridicos documentados, AJd)* is payable at between 0.5% and 1% (dependent on the region). VAT would be added to this figure as well. If you purchase the property with a Spanish mortgage, further *AJd* and VAT will be due on the amount of the mortgage in addition to the amount due on the value of the property.

Land registration fees

Finally, there are land registration fees which are about the same as the notary's fees, and are chargeable according to the property price.

If you purchase the property with a Spanish mortgage, this will also have to be registered, and a further charge will be payable for this.

Mortgage and other fees

If you are taking out a Spanish mortgage, you should also budget for the arrangement fee you will have to pay the Spanish lender. Spanish lenders normally charge a fixed arrangement fee of 1% of the mortgage amount. However, some lenders may 'cap' the amount payable to an agreed limit or may be willing to offer a discount if the amount you are borrowing is considerable.

You should also bear in mind the cost of your currency, and of any transfers of funds between the UK and Spain. A reputable currency broker will be able to advise you regarding this matter, and will probably be able to offer you a better rate of exchange then your own bank, and may even be able to transfer Euros to a Spanish bank account for less than your bank.

1.19 Insurance

Do not forget to insure the property and its contents. If you are using it as a holiday home (or intend to let it) ensure this is declared to the insurance company, as any failure to do so may invalidate the policy. You should also check the policy if it is for a holiday home, to see what risks are covered, as some insurance companies will not cover for, say, flood or fire if the property is unoccupied for more than 30 days at a time.

If you are intending to let out the property, this should also be declared, as again, failure to notify the insurance policy may invalidate the policy if damage occurs while the property is let.

1.20 Urbanisation problems

There is a law in Spain which only affects rural property and, until now, has only really been applied in Valencia, although it has been seen in a lesser way in Andalucía and Murcia. It allows a developer to put in an urbanisation proposal (that is to build a small development, usually on a subsidy), which will enable that developer to seize land for the development, or force the owner to pay cash towards the utilities required for that development. The owners may only have 15 days after notification to object, and of course, there is no guarantee that any objections would be accepted.

This law has been abused recently by developers to their own advantage.

There has been much published about the seizure of land in this way, and the government has promised an amendment to the law to prevent the current abuses. This is a controversial matter which has been ongoing for some time now, that has still not been resolved, and was still being debated at the time this book went to press and a pressure group – *Abusos Urbanisticos NO* – of more than 1,000 members, mainly expats, had been formed to campaign actively against such seizures.

"There is no such thing as a good tax!"

WINSTON CHURCHILL

Property ownership

Whether buying, selling or simply owning, property ownership brings significant tax implications in Spain and the UK. For advice on how to maximise your wealth and minimise your capital taxes, contact us to arrange an appointment.

We need to talk.
www.blevinsfranks.com

CHAPTER 2

Foreign Exchange and Overseas Property

2.1 What is foreign exchange?

The foreign exchange market is vast. Every day £1,000bn is traded through thousands of dealing rooms in dozens of countries throughout the world. Amazingly, this figure is greater than the global equity markets (shares) trade in an entire year. Of this amount, 95% is traded for speculative gain and is largely accounted for by banks trading with each other, known as 'interbank' trading. However, this still means that £50bn is traded daily for commercial reasons for both businesses and private investors alike.

The UK is by far the world's largest financial market. In fact, the UK trades a greater volume than the financial markets of New York and Hong Kong put together. This is partly due to the historical significance of London as a world trading centre and partly because of the time zone advantage the UK has over other countries, which enables it to cover both European and US hours of trade.

2.2 What relevance does this have with my overseas property purchase or move abroad?

More than you might think. The timing of when to buy or sell currency is at the heart of every foreign exchange transaction and with property deals taking anything up to 18 months and beyond, control and understanding are key to success.

Most people identify foreign exchange with the buying of holiday money, however, this bears little relation to the bulk buying of currency for a property transaction.

The annual buying of holiday money provides scant opportunity to identify with the speed exchange rates move, and the ranges within which they trade. This is often the most surprising aspect of foreign exchange for those unfamiliar with the markets.

However, when understood and correctly harnessed, exchange rate volatility and near 24-hour, seven-days-a-week trading opportunities can present excellent opportunities to save you money. Conversely, anyone choosing to ignore the market will either fail to make the most of these opportunities or, at worst, will find the cost of their property escalate beyond their budget.

Furthermore, those who are seeking to actually move to another country, such as Spain, will have to consider where they will keep any capital, and when and where to move any income into Euros, particularly when also keeping an eye on their cashflow. It is all very well stockpiling funds in the UK and transferring these into Euros when the rate is good, but what do you live on in the meantime? And what happens if the rate continues to move against you?

2.3 Will my bank help me transfer money abroad to purchase my property?

Over the years, banks have traditionally been the mainstay of foreign exchange and money transfer services. Consequently, they tend to be the first place people who are intending to buy overseas property turn to.

However, private client transactions are often viewed as one-off deals and, as such, are usually subject to uncompetitive exchange rates, which enable the banks to generate sufficient profit in their deals.

In addition, a transfer within Europe can often take three or four days and information flows are scarce and trading hours limited to branch opening times. There can also be costs incurred in transferring out of an account: the receiving bank may charge a fee, and there may also be commission fees on the currency exchange.

2.4 Foreign exchange, foreign language

With normal conveyancing periods of two months or more and the construction of newly built properties taking anything up to 18 months and beyond, a simple, information led foreign exchange plan can take much of the confusion and worry out of buying overseas.

Unfortunately, the banks will offer you limited choice. More often than not, they will simply advise you to buy your currency on the day that you need to transfer money. Not only is the exchange rate likely to be uncompetitive but you will also have missed out on any opportunities that the markets may have presented in the meantime. You may also suffer added expense if the exchange rate has moved against you, which may upset your budget.

Fortunately, more timely and readily available information flows, a single dedicated specialist point of contact, 24-hour exchange rate monitoring and simple solutions to achieve better, more commercial, rates of exchange are now not only the preserve of 'blue chip' organisations, and there are many foreign exchange brokers who will deal with private individuals.

2.5 Buying foreign currency

There are several different solutions to suit many different needs, each of which is outlined below.

Spot contract

This is an agreement to buy or sell currency for immediate delivery to a nominated bank account (up to five working days ahead).

Spot contracts are ideal for people who need to transfer money quickly, and enable you to buy foreign currency immediately. This is the simplest way to buy currency. If you need to make a fast payment, then this contract may suit you.

Forward contract

This contract offers you the opportunity to fix an exchange rate for up to two years in the future, while paying only a small deposit at the outset and paying the balance on maturity of the contract. Forward contracts can be for a set date in the future or can be flexible, depending on your needs. Forward contracts are ideal for people who need to make a payment in the future and wish to lock into a favourable exchange rate or to protect themselves from adverse exchange rate movements.

This fixed exchange rate enables you to calculate the Sterling cost of your property and, as it does not need to be paid for until the maturity date (although a deposit of around 10% of the total is usually due when the contract is entered into), it enables you to budget and resolve any immediate cashflow concerns. This is often the case if, for instance, your house sale has not been completed, yet you want to secure an advantageous exchange rate.

As the availability of your funds, the payment plan of your new property and the best available exchange rate rarely coincide, forward contracts are a popular option. They are a cost effective and simple way of securing an exchange rate and simultaneously eliminating exchange rate volatility.

Stop loss order

Exchange rates fluctuate very rapidly, and often the best rates appear and disappear in the blink of an eye, or often at inconvenient times. Different firms give very slightly different rates, and in the time taken to ring up two currency brokers, the rate can have changed again. Automated orders (orders placed whereby you set specific limits of the rates you wish to buy at) are very popular for those with more time, who can afford to wait for a spike in the exchange rate.

A stop loss order is an automated market order whereby you set a minimum limit at which, if the markets reach this level, you will buy or sell your currency. This effectively guarantees the minimum rate that you will get. Stop loss orders are suitable for people who have funds to transfer but who have no real time constraints and are therefore seeking to make the most of any favourable currency movements, while protecting themselves from any adverse currency movements.

Limit order

This is another automated market order whereby you set a target exchange rate at which, if the rate is achieved, you will buy or sell your currency. Limit orders are suitable for people who have funds to transfer but who have no real time constraints and are therefore looking to make the most of a favourable currency movement, if one occurs. Stop loss and limit orders can be run together to make the most of positive currency movements, while offering protection from negative movements.

Regular payment plan

If you are making regular currency transfers, such as overseas mortgage or moving UK pension payments abroad, you may wish to set up a regular payment plan.

This service removes the worry caused by exchange rate fluctuations when making periodic currency payments over a period of time.

Aside from the obvious benefits of receiving excellent exchange rates, which are fixed for a period of up to two years, the regular payment plan also offers you the following significant advantages:

- No commission charges.
- Cheaper transfer fees than the bank.
- Faster transfer times than the bank.
- The convenience of paying by direct debit.

Once you have agreed the amount of currency to be transferred each month, this cannot be changed. We therefore strongly advise you to consider any potential changes in your monthly requirement (eg, increases in mortgage payments) and make allowances for these when booking your payments.

It is also possible to combine some or all of these payment methods, especially for smaller, regular payments, eg, one pension income stream can be transferred monthly on a 'spot' basis – the amount will be irregular, depending on the exchange rate, but will be a constant source of income. Another income stream could be 'forward bought' when the exchange rate is advantageous, giving you a guaranteed level of income for a specified period, and again, cash-flow. Yet another stream could be stockpiled in the UK and transferred when the rate is beneficial.

2.6 How does this relate to my property purchase?

Budgeting your expenditure accurately is impossible without knowing what exchange rate you will get to convert your Sterling into foreign currency, although there may be a number of scenarios to consider, for instance:

- You have agreed to buy a house overseas and need to transfer your deposit there immediately.
- You need to make three additional staged payments over the next 12 months.
- You have retired overseas. You know that your pension, paid in Sterling, will be £15,000 per year. What will this be worth in the country where you are now living?
- You have agreed on a local currency mortgage to buy your overseas property and will be paying for it out of your Sterling income. How much will your mortgage payments cost over the next year?

SO, WHAT ARE YOUR OPTIONS?

Do nothing

This is the most common action from overseas property buyers and, sadly, it often costs them money. If the exchange rate for

Sterling against the local currency improves, people are happy. If it worsens, then they kick themselves.

The value of the Euro can and does fluctuate on a daily basis, and sometimes even across a range of a few cents a day. This volatility will ultimately affect the final price that you pay for your overseas property.

Buy the currency now

This will tie up large quantities of your money for long periods. The overseas funds will earn very little interest because continental interest rates are usually lower than UK rates.

'Forward' buy the currency

You can fix an exchange rate for delivery on an agreed date in the future. Money can be drawn down whenever you want it during the period of the contract or the contract can be extended if, for example, the construction of your property is delayed.

You will need to place a deposit of 10% of the amount you wish to buy, with the balance payable prior to the maturity of the contract.

Use a combination of the above methods

By looking at what you want to do and when, taking into account your specific circumstances, you can utilise any, some or all of these different tools to minimise your exposure to adverse currency rates, and maximise your chances of taking advantage of beneficial rates or spikes in the rates, however short the duration of these fluctuations.

2.7 Draw comfort from 24-hour market monitoring

Banks' restricted opening hours and the often limited knowledge of regional branch staff of the intricacies of foreign exchange may often leave you feeling short changed. Smaller foreign exchange specialists will monitor your transaction from dawn till dusk, enabling you to take advantage of any opportunities, which you may otherwise have not been aware of.

2.8 Risks

Whenever you enter into a foreign exchange transaction (even with your own bank) you have a delivery risk. This arises because you have to deliver your currency (say Sterling) before you receive the foreign currency (say Euro). This delay may only be for a few hours, or overnight, so the risk of non-delivery of your currency is just for a short time, especially in this electronic world. Some solicitors have arrangements whereby your money is only released to the exchange company once they have received written confirmation of the forwarding of the foreign currency to the designated beneficiary account. This eliminates risk.

CHAPTER 3

Spanish Tax Residence and Scope of Spanish Taxes

3.1 Introduction

You will become resident for tax purposes in Spain if:

- You spend more than 183 days there in one calendar year. These days do not have to be consecutive. You become liable whether or not you take out a formal residence permit. Temporary absences from Spain are ignored for the purpose of the 183-day rule unless it can be proved that the individual is habitually resident in another country for more than 183 days in a calendar year. Days of arrival and days of departure do count as days of presence in Spain; *or*
- Your 'centre of economic interests' is in Spain, ie, the base for your economic or professional activities is in Spain; *or*
- Your 'centre of vital interests' is in Spain – ie, your spouse lives in Spain and you are not legally separated, and/or your dependent minor children do. In this case you are presumed Spanish resident, unless proven otherwise, even though you may spend less than 183 days per year in Spain.

In Spain, split-year treatment does not apply. You are either resident or not resident for the whole tax year. A Spanish tax year is the same as a calendar year (1 January to 31 December), unlike the UK tax year which is from 6 April to the following 5 April.

If you are a resident of Spain you will be liable for income tax and capital gains tax on your worldwide income and gains, and wealth tax on your worldwide assets. This is in addition to other incidental tax liabilities such as *IVA* (VAT) and local taxes. Succession tax, which is a tax on inheritances and gifts, is payable if the recipient is resident in Spain, or the asset is situated in Spain. It is worth noting that succession tax in Spain is payable by the recipient of the gift or inheritance.

If you live on a boat within 12 nautical miles of Spanish land, you are a Spanish tax resident. A day within 12 nautical miles is a day spent in Spain for tax purposes.

3.2 The UK/Spain double tax treaty

The double tax treaty between the UK and Spain has a 'tie-breaker' clause that comes into operation if you are resident both in the UK under the UK rules and in Spain under the Spanish rules. The purpose is to determine in which country you will be regarded as resident for the purpose of taxes covered by the agreement – it cannot be both.

The agreement works as follows:

- If you are resident in both countries according to each country's domestic rules, you are deemed to be resident in the country in which you have a permanent home available to you. A permanent home is any form of accommodation which is continuously available to you for your personal use. It does not have to be owned by you.
- If you have permanent homes available to you in both countries, you are deemed to be resident in the country that is your centre of vital interests, ie, the country with which your personal and economic relations are the closest. 'Personal and economic relations' is a broad expression intended to cover the full range of social, domestic, financial, political and cultural links. The whole range of these factors is taken into account, but considerations based upon personal acts are given special weight. For example, if you have a home in the UK and set up another in Spain while retaining the first, the fact that you have retained your home in

the UK, where you have lived, worked, and where your family and possessions are, can, together with other factors, demonstrate that your personal and economic relations are closer to the UK than Spain. If this test is indeterminate, you are deemed to be resident in the country in which you have an habitual abode. This broadly means the country in which, over a reasonable period, you stay more frequently. The comparison must be made over a sufficient length of time for it to be possible to determine whether residence in either the UK or Spain has become 'habitual'.

- If you have an habitual abode in both countries, you are deemed to be resident in the country of which you are a national. UK nationals will at this point be regarded as UK residents.

It is dangerous to rely on a double tax treaty to make you resident in one country or the other; your interpretation of the treaty may not agree with that of one or both of the countries, or you may fall ill and inadvertently slip out of the treaty rules into the domestic rules of one of the countries. This could be an issue for you if you have not planned for the move.

3.3 Coming to/leaving Spain part way through the tax year

Under Spanish domestic laws, you are either resident or non-resident for the whole tax year.

So, if you move to Spain part way through the tax year, the date from which you become resident in Spain will largely depend on when you arrive in Spain.

If you arrive in Spain (with the intention of staying there indefinitely) during the first six months of the calendar year, you are likely to be regarded as resident in Spain for the full calendar year under the Spanish residency rules, as you will have spent more than 183 days in Spain during that year (short absences are ignored once you have settled in Spain). However, if you moved to Spain directly from the UK, then it is likely that, under the UK/Spain Tax Treaty rules, you will be regarded as UK resident up to the date you leave the UK and resident in Spain thereafter (although Spain will still not split the tax year).

If you move to Spain in the latter half of the calendar year, then you are likely to find that you are regarded as non-Spanish resident during that year, assuming you have not spent 183 days there during the year. However, this will depend on previous visits made to Spain in that year, and if these have been significant or frequent, the Spanish authorities could deem you to be tax resident in Spain from an earlier date, and may merely regard any subsequent time spent outside of Spain as a temporary absence.

The centre of interests rules were introduced to prevent fraud whereby individuals with all their personal, financial and economic ties in Spain were claiming non-Spanish residence on the basis of time spent outside Spain. Therefore, this rule is unlikely to apply in your year of arrival in Spain.

If you leave Spain part way through a tax year, the same principles will apply. In the first instance, your absence from Spain will be deemed to be temporary unless you can prove residency in another country. This is usually done by means of a tax certificate from the authorities in the new country.

So, if you leave Spain during the first half of the calendar year, and you move back to the UK, say, you will not be regarded as resident in Spain for that year, as you will not have spent 183 days in Spain during that year. This assumes you do not go back to Spain during that calendar year, as the intervening time spent in the UK could then be regarded as a temporary absence.

If you leave Spain during the second half of the calendar year, you will be regarded as resident in Spain for the whole of the year. However, if you move directly to the UK, for example, then it is likely that under the tax treaty rules you would be regarded as resident in Spain up until the date you leave, and resident in the UK thereafter.

3.4 Leaving Spain for a tax haven country

Where a person of Spanish nationality moves to a tax haven, they are treated as a continuing resident in Spain for income and capital gains tax purposes for the year of their move and the following four tax years.

3.5 Fiscal identity number *(numero de identificacion: NIE)*

If you reside in Spain, or if you own a property and are non-resident, you need a *número de identificación* or *NIE.* Most significant property transactions require you to provide this fiscal identification number. The number identifies you to the Spanish tax authority and is required when you pay your taxes or have any dealings with the *Hacienda.*

Previously, it was possible to buy Spanish property prior to obtaining this number, however changes in the law now require all foreigners buying property to secure their *NIE* and the accompanying letter before completing their purchase. An *NIE* is also highly important for any expat living in Spain as it is needed when: setting up your utilities, applying for a Spanish driving license or validating an EU one, applying for a business permit, any dealings with the tax authority, inheriting any Spanish assets and making purchases above €3,000.

To obtain one, present yourself at the nearest police station or *comisaría* (with a foreigners' department) with the following documentation:

- A completed and signed application form and a photocopy of the completed form.
- Your passport (and a photocopy).
- Proof of address in both the UK and Spain (if you do not have a Spanish address, that of a friend, solicitor or estate agent will do).
- Written justification of why you need the *NIE.* This can be issued by an accountant, notary, solicitor, estate agent, bank manager, insurance agent, former employer, etc.

The *NIE* will take two to six weeks to come through.

There are three types of Spanish tax identification numbers:

1. *CIF* for companies.
2. *NIF* for Spanish nationals.
3. *NIE* for foreign residents or foreigners wishing to buy property in Spain.

3.6 Tax returns

Individuals are required to return a completed tax form (obtainable from *estancos* – tobacco kiosks) normally in May or June (up to 20 June) of each year in relation to income and capital gains made during the preceding year ending 31 December. While individuals with simple tax affairs do not have to submit an income tax return provided certain conditions are met, submission of a return can be useful in proving that you have been resident in Spain when selling the property and claiming relief against the gain, for example. However, on the face of it, it would appear that in many cases these people would be entitled to a tax refund if they completed a Spanish tax return. So it is worthwhile for a Spanish tax resident to complete the return even if the exemption applies.

Tax may be paid in two instalments with the second payment being due, in general, in the following November.

Every year, the Spanish Government changes the tax tables and exemptions, and in some years it decides to radically change the rules.

3.7 Husband and wife can have different tax residences

This is possible, but advice specific to your circumstances should be taken.

3.8 Temporary absences from Spain: a trap

If you arrive in Spain (or live there) and then later leave for a temporary period (eg, a few months) the Spanish tax authorities will usually deem you to be Spanish tax resident throughout your absence.

Days of arrival and days of departure do count as days in Spain, as well as temporary absences (eg, short periods spent outside Spain).

3.9 Fiscal nomads

It is possible not to be a resident anywhere in the world, though advice should be taken on how exactly to achieve this. Such a person is called a 'fiscal nomad' or a 'permanent traveller'. However, income usually remains taxable in the country where it originates in this case, although if you are a fiscal nomad then certain UK income can be UK tax-free as long as you do not claim a UK personal allowance.

3.10 Scope of Spanish taxes

	Spanish Tax Resident	**Not Spanish Tax Resident**
Income Tax	Worldwide – at scale rates and after allowances and deductions not available to non-residents Investment income at 21%, 25% and 27%	Only on Spanish source income – but generally at fixed rates with no allowances or deductions
Capital Gains		
Own Home	Usually tax-free (see notes)	Spanish property taxable at 21%, 25% and 27%
Other Property	Worldwide at 21%, 25% and 27% Short-term gains (asset held for less than 12 months) are taxed at the normal scale rates	Spanish property taxable at 21%, 25% and 27%
Wealth Tax	Worldwide with allowances	Only on Spanish property
Succession Tax	Worldwide almost (see notes)	Only Spanish property
Local Taxes	On Spanish real estate	On Spanish real estate

CHAPTER 4

Spanish Income Tax

4.1 Tax rates

In Spain, income is split into general income *(renta general)* and savings income *(renta del ahorro)*. Savings income includes capital gains. After being calculated according to the rules for each particular type of income within each category, the total of general and savings income is termed the *base imponible* (or taxable base). After any deductions and allowances it is then known as the *base liquidable* (net taxable base).

In 2012, Spain introduced 'temporary' tax increases to both savings and general income. These were originally only intended to apply for 2012 and 2013 income and gains, but were extended to apply to 2014 as well.

Savings income

The worldwide savings income of residents and the Spanish savings income of non-residents (subject to tax treaty) was taxed at a fixed 19% on the first €6,000 and 21% on income over that amount.

Following the temporary tax rises which were introduced for 2012, 2013 and 2014, the tax rates on savings income have been increased to:

Up to €6,000	21%
€6,000 to €24,000	25%
Over €24,000	27%

Savings income basically consists of:

- Dividend income.
- Interest income.
- Income derived from life assurance contracts.
- Purchased annuity income.
- Capital gains on the sale/transfer of assets.

General income

Spanish residents are taxed on their worldwide 'general' income at progressive scale rates. The income tax scale rates are made up of the 'State tax rates' and the 'Community tax rates' (which are set independently by the particular Autonomous Community).

Generally, the top combined rate of tax is 45% (52% under the 'temporary' tax rises mentioned above), but can be higher in some Communities if they have increased the Community tax rates applying in that region. For example, in Cataluña and Andalucía it is 56%.

Anything not categorised as 'savings income' is included as 'general income', including all earned income (ie, salary, self-employment and pension income), rental income, any imputed income and gains not made on the sale/transfer of assets such as from gambling, for example.

The income tax rates for 2012, 2013 (and proposed for 2014) are:

From €	To €	National Income Tax Rate %	Community Income Tax Rate %	Total Tax Rate %	Tax Payable on Band €	Cumulative Tax Payable €
0.00	17,707.20	12.75	12	24.75	4,383	4,383
17,707.20	33,007.20	16	14	30	4,590	8,973
33,007.20	53,407.20	21.5	18.5	40	8,160	17,133
53,407.20	120,000.20	25.5	21.5	47	31,299	48,432
120,000.20	175,000.20	27.5	21.5	49	26,950	75,382
175,000.20	300,000.20	29.5	21.5	51	63,750	139,132
300,000.20	–	30.5	21.5	52	–	–

In Andalucía, the 2013 rates are:

From €	To €	National Income Tax Rate %	Community Income Tax Rate %	Total Tax Rate %	Tax Payable on Band €	Cumulative Tax Payable €
0.00	17,707	12.75	12	24.75	4,383	4,383
17,707	33,007	16	14	30	4,590	8,973
33,007	53,407	21.5	18.5	40	8,160	17,133
53,407	80,000	25.5	21.5	47	12,499	29,632
80,000	100,000	25.5	22.5	48	9,600	39,232
100,000	120,000	25.5	23.5	49	9,800	49,032
120,000	175,000	27.5	25.5	53	29,150	78,182
175,000	300,000	29.5	25.5	55	68,750	146,932
300,000+	–	30.5	25.5	56	–	–

Spouses or unmarried couples with a *'pareja de hecho'* or *'unión de hecho'* can decide whether to complete individual returns or a joint return, and will usually calculate which of the two results in the payment of less tax or the receipt of the highest tax refund, before making that decision.

Non-residents pay tax on general income at a flat rate of 24.75% and on savings income and gains at the same rates as residents.

However, those non-residents resident in the EU receive the same tax treatment as that applicable to residents, so can, for example, deduct the same expenses as residents when calculating taxable rental income.

4.2 Tax threshold: when do you need to complete a tax return?

The rules about declaration are complex. However, there appear to be two fundamental principles. Firstly, the Spanish tax authorities want a Spanish tax resident to complete a Spanish tax return if they have complicated affairs, particularly if they have quite a significant amount of 'unearned' or investment income. Secondly, the Spanish tax authorities allow individuals who have already had tax deducted (for example on their salaries) not to complete a Spanish tax return if they have little other income and do not wish to do so. It is worth noting that if many people in this position did complete a tax return they would probably receive a refund.

You do not have to complete a Spanish tax return even though you may be Spanish tax resident if:

- Your income is from one source only, it is earned income (which includes pensions), it is less than €22,000 and tax has been deducted at source in Spain. This exemption does not therefore apply to overseas pensions since these do not suffer Spanish withholding tax.
- Your income is from more than one source, or where there is no withholding tax at source, and is no more than €11,200 in the year. So if you have an overseas pension of less than €11,200, and it is your only income, you do not have to complete a tax return.

- Your income from both investment income and capital gains is no more than €1,600 gross in the year, but subject to withholding tax at source.
- Imputed rental income is no more than €1,000 per year.

Note that these limits are for individual tax returns. If you decide on a joint return, the same limits apply (there is no increase in the limit).

No return is required if your total income from all sources is less than €1,000. If, however, you want to claim any tax breaks, you do have to file a tax return.

For example if:

- You are entitled to a Spanish tax break related to housing (ie, you have a mortgage or a special Spanish tax-qualified savings account to buy a house).
- You are entitled to international double tax relief (ie, you have income subject to tax at source in an overseas country).
- You are entitled to tax relief for payments into a pension fund or similar scheme.

4.3 Deductions

You can deduct:

- All payments into the Spanish social security system.
- Your contributions to a qualifying pension scheme (qualifying, that is, for Spanish income tax purposes) up to certain limits.
- Alimony/Maintenance payments under a court order to a former spouse, but not payments for the benefit of a child. (The ex-spouse's income is taxable in his/her hands but maintenance for a child is not.) Payments made by a UK resident to his/her Spanish resident former spouse will receive little, if any, relief in the UK and yet will be taxable income in Spain in the hands of the recipient.

Earned income allowance (including pensions)

The recipient of earned income (including pension income – but not, say, rental income) has a deduction of up to €4,080 from that earned income. The maximum amount of €4,080 applies to those on low earned income of up to €9,180. The minimum deduction for those who earn more than €13,260 is €2,652. If earning some figure between €9,180 and €13,260, the deduction is a figure between €2,652 and €4,080, calculated using a formula: ((a) €4,080 less (b) 35% of (c) the actual figure less €9,180 – ie, (a) – [(b) x (c)]). So, if earning €10,000 in the year, the deduction would be €3,793 and the net figure taken into account for tax purposes would be €6,207.

These deductions are higher for handicapped workers (€3,264 for 33% to 65% incapacity and €7,242 for incapacity above that).

- Those working beyond the age of 65 receive twice the allowance, as do those who move to a new area to take up work having been unemployed. This increase is usually available in both the year of relocation and also the following year.

Payments to political parties

You can deduct payments made to political parties, federations, coalitions, or other such groups, up to a maximum of €600 per year.

Once these deductions have been made from your income, the amount left over (your net taxable income or *base liquidable*) is taxed at the progressive rates set out above. These deductions cannot result in a net negative income figure.

There are other kinds of deductions, called tax credits, which are deducted not from the taxable income but from the amount of tax payable, calculated using the sliding scale rates noted above. Perhaps the most important are deductions relating to mortgage payments on your main Spanish home (see below), and there are, of course, final deductions for previous payment of Spanish taxes and for payment of overseas taxes if you have paid these, or had them deducted – (for example UK tax deducted from your UK income if you have any).

There are a few local variations in Spanish income tax deductions depending on the area of Spain ie, the *Comunidad Autónoma*, in which the Spanish taxpayer is registered. The local variations are clear in the national Spanish tax forms, and are not very dramatic.

4.4 Tax credits

Purchase of main residence

There is a tax deduction of 15% of the amounts paid in the year up to €9,040 for purchase or improvement of the family home – ie, a maximum tax deduction of €1,356. It includes additional construction costs but not general improvements or repairs, or garages. It applies to any actual outlay, mortgage repayments and interest, and also to amounts paid into designated 'homebuyer' accounts for future purchase of a main residence in Spain.

For properties purchased prior to 20 January 2006, so long as at least 50% of the value of the property is purchased with arm's length finance and the repayments over the first three years do not repay more than 40% of the loan capital, this 15% deduction can be increased. For the first two years you benefit from a higher deduction of 25% on the first €4,508 per year and then 15% of the excess up to €9,015; thereafter, 20% of the first €4,508 and 15% of the excess up to €9,015.

Renting your main home

You can deduct 10.05% of the amounts paid on renting your main home, provided your taxable income is less than €24,107 per year.

The maximum deduction for 2013 and 2014 is:

- €9,040 where your taxable income is equal to or less than €17,707.
- €9,040 less the result of multiplying by 1.4125 the difference between your taxable income and €17,707, where your taxable income is between €17,707 and €24,107 per year.

The *'Minimo Personal y Familiar'*

This is a deduction against the calculated tax liability, and although it is a tax credit, not a deduction against income, is the closest thing the Spanish have to a personal allowance. Any allowance not used against the 'general' income can be set against the 'savings' income. The allowance consists of the total of:

- The basic allowance for the taxpayer (including age additions) and further allowances for;
- Dependent children/grandchildren (descendants) living with the taxpayer; *or*
- Dependent elderly or disabled parents/grandparents (ascendants) living with the taxpayer; *and*
- For the incapacity of the taxpayer or any descendant/ascendant living with the taxpayer.

In some circumstances considerable tax-free allowances can be available to a household, especially where there are dependants and/or incapacity.

Basic allowance

The basic personal allowance for 2013 is €5,151 per person (this has not changed for many years). For joint returns the allowance given to the first spouse is €5,151 plus €3,400 for the second spouse. A single parent receives €5,661.

This is set against the lowest rate of tax payable, so 24.75% for general income, and 21% for savings income.

Curiously, the second spouse allowance (€3,400) is still given as a deduction from taxable income, and not as a tax credit.

Age additions

The basic allowance is increased by €918 for someone aged 65 or more (so €6,069 in total) and by a further €1,122 for someone aged 75 or more (€6,273 in total).

Dependents

The additional amounts for dependent children (descendants) or aged/disabled parents (ascendants) living with the claimant are:

- Child under 25 (or disabled) and single: first €1,836, second €2,040, third €3,672, fourth, fifth, etc, child €4,182.

NOTE

Where the child is younger than three years old, an extra €2,244 is available per child.

- Parent or other ascendant aged 65 and over (or disabled): €918 per dependant.
- Parent or other ascendant aged 75 and over: further €1,122 (ie, €2,040 in total) per dependant.

To obtain the family deduction the additional family members must be living in the taxpayer's home and their own income must total less than €8,000 per annum (excluding exempt income).

Disability

Disabled taxpayers and dependent relatives as above: €2,316 (more than 33% disability) or €7,038 (more than 65% disability) in addition to any allowances above.

The allowance is increased by a further €2,316 for assistance costs where someone needs the help of a third person, or is of reduced mobility, or 65% disabled.

The increase in the personal allowances depends on the 'grade' of the disability. Where an individual is in receipt of incapacity benefit in Spain due to a permanent incapacity to work, the grade of disability is classed as 33%, and the personal allowances are increased by the €2,316 mentioned (and more depending on the circumstances). You would normally be required to have a certificate from the relevant social security authorities in Spain *(IMSERSO)* confirming the grade of disability. It is not clear whether the increased personal allowance applies to someone in receipt of UK long-term incapacity benefit (now Employment and Support Allowance – ESA) and this should be checked locally.

The amount of tax credit available is calculated by taking the total personal allowances available and multiplying this figure by the scale rates of tax, starting with the lowest rate first. The tax credit thus calculated is then deducted from the taxpayer's total tax payable for the year.

This has the effect of relieving tax not at the taxpayer's marginal rate, but first at the lowest rate of 24.75%, and then at the higher rates where applicable (eg, where the taxpayer has particularly large allowances in total because of age, dependants, incapacity, etc).

For an individual with the basic allowance of €5,151, the tax credit is worth €1,275 (ie, €5,151 x 24.75%). This amount is deducted from the individual's Spanish tax liability for that year.

The credit is relieved first against the tax payable on income taxable as general income. If it is not fully utilised, the unused amount of allowance (not amount of credit) is then deducted from the amount of income taxable as savings income.

4.5 Taxation of rental income

The following applies to all rental income from furnished or unfurnished lettings:

Residents of Spain

For 2013, a reduction of 60% is available against net rental income for residents of Spain, before tax is payable. This includes any lettings income from outside Spain but **not** short-term holiday lets *(por temporada)*. There is no specific time limit past which a contract becomes long-term, but in Spain the lease would normally be at least one year in duration and be potentially renewable at the tenants request for up to five years. Long-term lets are where the tenant truly makes the apartment his home, although it does not have to be his principal residence. In Spain, long-term tenants have rights that short-term tenants do not have. Short-term holiday lets are designed for holiday rentals and may only last a matter of months at most and are not automatically renewable when the contract ends. A similar distinction between short and longer term lets can be made for, say, UK rental income.

In certain circumstances, if a Spanish property is let on a long-term basis to young people aged between 18 to 35, up to 100% of the rental income may be exempt. From 2011, the maximum age of the tenant was reduced to 30 for the 100% exemption to apply. However, landlords who rented to an individual under 35 before 31 December 2010 will retain the exemption until the tenant reaches 35.

The net rental income is the amount of rent due after deducting usual day-to-day running costs for the period(s) in question, including local taxes such as *IBI* (the 'rates'), repairs and maintenance (but not additions or improvements/enhancements that add to the value of the property), managing agents' fees and commissions, interest on loans for purchase or improvement, and depreciation of 3% per year of the cost of the property (excluding the land value).

Repairs and maintenance expenses – income vs capital expenditure

Repairs and maintenance expenses include repairs to 'maintain the normal use of the item', and also where there is an element of 'replacement'. So, where central heating, lifts, security doors, etc, were not there before and are subsequently installed, these expenses are not repairs but improvement expenses allowable against the capital gain. However, where these items already existed, and it is necessary to replace them for the normal use to continue, the costs are allowable as a revenue expense. Replacing a coal heater for an oil heater is an improvement expense, not a repair.

The renovation of an obsolete and potentially unsafe electrical installation is a repair. On the other hand, an electrical installation necessary for a new room in a property is an improvement.

Expenses incurred with the aim of maintaining the property in its original state are treated as repairs, provided there is no change to the structure or surface area of the property. Anything which maintains the useful life of the property or its productivity is allowable as a revenue expense. Expenses to increase productivity or prolong the useful life of the property are capital expenses and allowable against the capital gain.

Mortgage interest

Spanish mortgage interest is allowable provided the mortgage was used to acquire or improve the let property.

Losses

An overall net negative rental loss can arise, but certain expenses are limited to no more than the total rental figure for the property in question. In this case the excess (loss) can be carried forward for up to the next four years against future profits from the same property. This applies to loan interest, other finance charges and repair and maintenance costs. It does not apply to other costs such as local taxes, utility bills, bad debts and depreciation.

Although there might appear to be an unrelieved loss attributable to allowable items when all expenditure is taken into account, it is only where the restricted allowable items exceed the gross income from a particular property for the year that an allowable loss can arise for carry forward.

For example:

- Gross income €1,000; loss-restricted expenses €800; other expenses €300.
- Current year net loss after all expenditure = (€100) taxable profit = nil.
- Net result after only 'restricted items' = €200 notional profit.
- Allowable loss to carry forward = nil.

Tax retention at source

The tenant is generally required to retain 15% of the rents on account of tax and pay this over to the authorities. This will be

24.75% in the case of a non-resident landlord (see below). In the case of short-term holiday lets it would be the agent or landlord who was responsible for this.

Non-residents of Spain

A non-resident is taxed in Spain on income arising from Spanish property at the rate of 24.75%. EU residents can deduct allowable expenses and are therefore taxed on the net rental income. Non-EU residents are taxed on the gross income.

The rental income remains liable to UK taxation if the landlord is UK resident but any Spanish tax paid can be deducted against any UK tax liability to avoid double tax on the income. In the UK, the taxable income will be calculated under UK rules, and therefore, deductions can be made for expenses wholly and exclusively related to the letting, including interest on any loan used to acquire the property.

It will be necessary to split the figures into UK tax years.

Notional rental income

If you own a property in Spain that is not used as your main home, a purely notional or theoretical rental income is deemed to arise for periods where the property is not actually let, based (normally) on 2% of the official value *(valor catastral)* of the property as shown in the *IBI* notice for the year. This rate drops to 1.1% of the *valor catastral* if it has been revised from 1 January 1994 onwards. The 1.1% rate also applies where a property lacks a *valor catastral* or it has yet to be notified. A Spanish property cannot be your main home if you are not officially resident in Spain, so this will always apply to property in Spain owned by non-residents, and to second, third, and more, properties owned by residents of Spain.

Where such a property is empty for part of the year and rented for part of the year, you calculate the notional income for the part of the year the property is empty.

No notional income is imputed where the property is continuously rented out (there does not have to be a minimum rent). Where, for example, there is a small gap between rentals while looking for another tenant, these gap periods can be ignored. What constitutes 'small' is open to debate and does not seem to be defined. If there are no lettings during the year, the notional rental income will apply for the whole year.

The notional income is simply added to your other income and taxed at the appropriate rate(s) – so for a non-resident owner it will be taxed at a flat 24.75%.

Non-resident taxpayers whose only asset in Spain is a Spanish property need to declare the deemed income tax on Form 214.

Where the property is subject to a usufruct, it is the usufruct holder who is liable to pay this tax. The tax is calculated on the full value of the property, not just the value of the usufruct.

A similar notional income rule applies where a property is let to a relative, but in this case the notional figure becomes the minimum taxable amount that can apply.

VAT *(IVA)*

Spanish VAT *(IVA – impuesto sobre el valor añadido)*, or *IGIC (Impuesto General Indirecto Canario)* in the Canaries, only applies to residential rentals where an active business approach is taken and additional hotel-like services are provided such as catering (eg, breakfast and/or other meals), cleaning and laundering. In such cases the business is also likely to be treated as a self-employment activity with all that this entails (see below).

Lettings of commercial premises are subject to VAT.

If VAT applies, it is at the reduced rate of 10% (or *IGIC* of 5% in the Canaries).

4.6 Self-employment income

Self-employment income is subject to tax at the progressive scale rates in Spain.

The amount taken into account for tax is the net profit: the gross income less the expenses of the business.

Social security contributions are also payable, and these are tax deductible in Spain. A minimum contribution of around €225 per month is payable, regardless of profits.

All self-employed traders and businesses *(autónomos)* must register for VAT *(IVA)* in Spain, regardless of turnover. The standard rate of *IVA* is 21% in Spain, although the letting of residential private property is normally exempt, as are health care, educational and certain other services.

Withholding tax and the self-employed

Where a self-employed individual who is resident in Spain is paid by a Spanish entity (or permanent establishment of a non-Spanish company), the Spanish entity has to withhold tax at a rate of 15% from the payments made to the individual. A reduced rate of withholding can apply during the first three years. It is up to the individual to inform the company/ies in writing that they have just started work and the company/ies must keep the notification as proof of exemption from the obligation to withhold at 15%.

Where a self-employed individual is paid by a non-Spanish resident company which does not have a permanent establishment in Spain, then no withholding tax needs to be applied.

Registration as an *Autónomo* (self-employed individual)

Certain steps have to be taken:

- Obtain your *NIE* number if you do not already have one.
- Apply for a tax licence for the activity to be carried out. This can be obtained from the *suma* office in the town where your activity will be centered, or Town Hall *(ayuntamiento)* as appropriate.
- Complete and submit a Form 037. This should be stamped by the tax office and confirms the method of payment of your taxes. There are two methods of paying tax:

1. You make quarterly VAT and income tax declarations using the 'direct estimation method', which means you have to start book-keeping, or use the services of an accountant.
2. You pay a quarterly fixed amount to cover income tax and VAT under the *módulos* system.

Both systems have their advantages:

Method 1 is a good method if you think your income is going to fluctuate greatly, as income tax and VAT are only paid on your actual profits/earnings.

Method 2 does not take into consideration whether you have had a good or bad month, as you pay the same fixed amount even if you have zero income one month. However, under this system, there is no need to prepare accounts or numbered VAT invoices. Once your business is established and turnover increases, you may well benefit from this method.

- In order to register for Spanish social security, you will need to complete and submit a registration document. You will then be liable for fixed monthly payments, which must be paid even if you have zero income.

  You will receive a temporary card that shows your social security number. You should take this card to your local social security clinic and register with a doctor. This will instigate your permanent health card.

There are various levels of social security contributions, depending on the amount of pension you wish to receive on retirement. Most people choose the minimum (currently just over €225 per month). At each level, you have the choice of paying an additional amount for temporary incapacity sickness benefit in the unfortunate event of your falling ill. You can change the option to pay this if you wish, but this must be done prior to 1 October each year. Those persons over the age of 50 have a slightly different payment structure, and workers over the age of 65 do not have to make any social security payments save for the optional temporary incapacity sickness payment mentioned above, although they can opt to continue paying if they wish.

4.7 Spanish business structures

Sole trader

This does tend to be an efficient form of operation for smaller businesses, and is the simplest form of business to establish, as it has no separate legal identity.

Sole traders are subject to Spanish income tax on the net income of the business (the gross income of the business less the expenses relating to the business). The net income will be added to any other income and taxed at the progressive scale rates.

A sole trader must also register for social security. The contributions are expensive, starting at around €225 per person per month.

There may also be requirements to secure certain licences and register for local business taxes on the business activities.

Co-ownership

This is a business that belongs to two or more proprietors but which is not as formal as trading as a partnership in Spain. The business has no separate legal identity, and therefore the bureaucratic processes involved in setting up the business are minimal. Each individual within the business will pay income tax and social security on their separate earnings, although *IVA* will be declarable by the whole entity, rather than the individual.

As with the sole trader, there are certain registration requirements and the business itself must obtain a tax identification number.

Partnership

In Spain, a partnership is an independent legal entity and therefore is more bureaucratically burdensome to set up than either of the previous options. A partnership is liable to pay corporation tax in Spain, rather than each partner being liable to individual income tax on their share of the income.

There is usually a minimum level of capital that must be invested in order to set up the partnership, and each individual partner will be subject to tax and social security on their share of the profits extracted from the partnership. The partnership is liable to *IVA* as an entity.

Company

This is administratively the most expensive and difficult form of business to establish, as Spanish companies cannot be bought off the shelf in the same way as UK companies. There are different types of company in Spain, and advice should always be sought if this is the preferred method of operation.

Employer social security contributions may be quite high, but these are deductible for the purposes of computing the taxable profit. There is a minimum wage in Spain.

If no profit is extracted by way of salary, there will be no social security liability on either the company or the individual (although in this case, if you run the company, and are not otherwise exempt, you may have to pay for private health care insurance). Both salary and profits extracted by way of dividends are subject to Spanish income tax on the recipient at the progressive scale rates of tax (although salary payments are tax-deductible in respect of corporation tax, as are employer social security payments, whereas dividends are paid out of profits already taxed).

4.8 Employment income

40% deduction

A 40% deduction is available for non-regular income generated over a period of more than two years, but from 1 January 2011 the deduction is not available where the non-regular income is over €300,000 (reduced from €600,000 during 2010).

UK/Spain tax treaty

Under the terms of the UK/Spain double taxation treaty, income arising from an employment exercised in one country by a resident of the other country is taxable only in the country of residence unless the employment is exercised in the other country. In this case, the individual is subject to tax in the country where the duties are performed on the proportion of the employment income relating to the duties performed there. Therefore, a resident of Spain working for and paid by a UK employer will usually be taxable in the UK on employment income relating to duties performed in the UK. If no duties are performed in the UK, then the earnings cannot be subject to UK tax. If all duties are performed in Spain, the earnings will be subject to tax in Spain.

Where an individual resident in Spain performs duties in the UK, then the earnings will be exempt from UK tax provided all the following conditions are fulfilled:

- The individual spends less than 183 days in the UK during the UK tax year (this includes part-days); *and*
- The remuneration is paid by an employer who is not resident in the UK; *and*
- The remuneration is not borne by a permanent establishment or a fixed base that the Spanish employer has in the UK.

Therefore, where a Spanish resident individual is employed by a Spanish company and all his remuneration is paid and ultimately borne by the Spanish company, no tax will be payable in the UK on any earnings attributable to UK duties, provided he spends less than 183 days in the UK during the tax year in question. Tax will be payable in Spain on his earnings.

Similarly, if a UK resident is employed by a UK company and all his remuneration is borne by the UK company, then even if he works in Spain, no tax is payable in Spain, provided he spends less than 183 days in Spain during the Spanish tax year in question. His entire remuneration will be taxable in the UK.

If you have employment income which falls to be taxed in Spain but you are not resident in Spain, the Spanish tax authorities will tax the gross income at a flat rate of 24.75%. You will not be entitled to the deductions available to Spanish residents on their employment income.

The European Commission considers this to be discriminatory and have requested that Spain review the tax rates applicable to non-residents of Spain on employment income. If Spain does not comply with the request, the Commission has the right to take the matter to the Court of Justice. If Spain does not comply within the two months allotted, and the case is taken to the Court of Justice, this may take some years to resolve.

Where an individual resident in Spain earns income in respect of work performed outside Spain for a non-Spanish employer, and the income arises in an overseas country which has income tax (ie, is not a tax haven) it is exempt from Spanish income tax up to €60,101.21 per year.

Termination payments

Termination or severance payments *(indemnizaciones)* are exempt from tax up to the limits set out in the workers' statute *(Estatuto de los Trabajadores)*. After that point, any further money received is taxable.

Stock options

Stock options are taxed at exercise on the difference between the exercise price and the fair market value at the time of exercise. The income is also subject to social security contributions in Spain.

The 'gain' may be tax exempt in Spain if certain conditions are met, and if not, a deduction of up to 40% may be available if the income 'non-regular' income.

Accommodation benefit

If this is paid to the employee in cash, then the amount paid as cash will be taxable in Spain (either at scale rates or at 24.75% if under Beckham rules – see below).

If the employer pays the accommodation directly to the owner of the property, the benefit is calculated at 10% of the *catastral* value of the property and is limited to 10% of the employee's remuneration. Where the property's *catastral* value has been reviewed after January 1994, the rate is 5%. The figure calculated in this way will usually be less than the actual rent paid.

It is important that the employee, and not the employer, signs the rental contract. If the contract is signed by the employer, VAT would become payable on the rents. If the employee signs the rental agreement, then no VAT is payable.

The 10%/5% restrictions do not apply where the individual is a director or shareholder of the company.

Benefits in kind provided to directors or shareholders

Directors and shareholders are deemed to be connected with the company under the Spanish 'connected operations' rules. This means that where a benefit is provided to such individuals, it should be valued at its market value.

Accommodation benefit, for example, will be valued at the market rent the property could be expected to fetch. In order to avoid the benefit in kind tax charge, the individual would need to pay a market rent for his or her occupation of the property. This benefit can apply where any company owns a property that is used by a Spanish resident – including the main home.

The value for medical insurance premiums and a company phone will be the amounts paid by the company to the insurance and mobile phone providers respectively.

A company car would be valued at 20% of its market value as if new.

Usually, where a director or employee receives a benefit in kind, it is taxed as employment income at the scale rates. However, where there is no employment relationship and the individual is not remunerated by the company, the benefits should be treated as income from moveable capital (arising from the shareholding in the company) and taxed at 21% (on the first €6,000 of total investment income), 25% on the next €18,000 and 27% thereafter. This is likely to apply when a property is owned by a company (and that is the sole asset of the company) and is used or occupied by a shareholder.

Taxation of expats resident in Spain: the Beckham Clause

This is the option whereby individuals who become resident in Spain as a consequence of an assignment to Spain can opt to be taxed as non-residents for the year of arrival and the following five years. The election must be made during the tax year when the employee becomes tax resident in Spain, and the individual will have to make an application to be treated as a non-resident.

Employees seconded to Spain after 31 December 2009 with annual expected employment income, derived from the international assignment, of more than €600,000 will be disqualified from applying for this regime. For the regime to be available, this limit must not be exceeded for each tax year in which the Beckham regime applies.

Employees assigned to Spain before 1 January 2010 will not be subject to this limit, and will continue to be eligible for the regime regardless of the level of employment income derived from the assignment.

The following conditions must be fulfilled:

- The individual must not have been resident in Spain at any time during the 10 years preceding their assignment to Spain.
- The assignment to Spain must be a consequence of an employment contract.
- The duties of the employment must be carried out in Spain.
- The work must be carried out for the benefit of a Spanish resident company or a Spanish permanent establishment of an overseas company. This is the condition that the Spanish authorities will scrutinise the most.
- The income derived from the employment must not be exempt from income tax in Spain under the rules for non-residents.

If the above conditions are fulfilled, the individual will be subject to a flat 24.75% tax rate on their employment income instead of the progressive scale rates. Personal or family exemptions do not apply under non-resident tax rules.

For all other tax purposes, the individual is treated as a non-resident of Spain. If the duties of the employment are undertaken in the UK, the income attributable to such duties may be taxable in the UK. Other countries have similar rules. Therefore, the 24.75% tax rate is likely only to be effective where the individual is not only Spanish resident, but also performs the duties (or by far the majority of them) in Spain.

To make the election, an application form must be filed with the Spanish authorities within six months from the commencement of the expatriate's Spanish activities. This is taken to mean within six months of the date the employee joins the Spanish social security system, or within six months of the date shown on Form A1 (formerly E101) if the employee remains in their home country social security system. The election cannot be made beforehand. A taxpayer who has elected for the new regime can withdraw their election in November or December of any tax year by means of a communication to their employers and to the Spanish tax authorities. The expatriate regime will then no longer apply to the individual in the following year. Once withdrawn, the election cannot be resubmitted.

Furthermore, the election form must be accompanied by certain supporting documents. The documents include a declaration from the employer confirming that the taxpayer is an employee and a copy of the Spanish social security registration documents or certificate of coverage if the individual is entitled to remain in their home country social security system. It is assumed that a copy of the assignment letter for international assignees or the employment contract for local employees will also be required. On correct filing of the election form and the above documentation the Spanish authorities must agree to the application of the regime in writing within 10 days. Those taxpayers electing for this regime will still maintain the right to request and receive a certificate of residence from the Spanish authorities, assuming that they meet the residence criteria established under the applicable double tax treaty.

4.9 Lottery winnings and premium bonds

Winnings from the Spanish National Lotteries (of which there are various types), including the Spanish Red Cross Lottery are taxable in Spain at a flat rate of 20% (withheld at source where the prize exceeds €2,500).

However, there is no exemption from Spanish income tax on winnings from other countries, such as the UK National Lottery or UK Premium Bonds. The latter are indeed taxable in Spain. The UK National Lottery is, technically speaking, only open to residents of the UK.

4.10 UK source income and UK tax for non-UK residents

For non-UK residents with UK source income, UK income tax on certain types of UK source income is limited to the combined amounts of:

- Tax deducted from, or treated as deducted from, or tax credits on 'disregarded income'; *and*
- The tax liability leaving out the disregarded income and with no personal allowances or double tax relief taken into account.

'Disregarded income' includes dividends and stock dividends from UK resident companies, interest, purchased life annuity payments, deeply discounted securities, distributions from unauthorised unit trusts, transactions in deposits, annual payments, the state pension, other social security pensions and taxable social security income, and certain payments made through UK brokers.

Therefore, there are effectively two ways in which a non-UK resident's UK source income can be taxed:

Usual Tax Treatment with personal allowance

All UK source income is taxed in the usual way after deduction of the individual's personal allowance.

Disregarded Income Treatment – no personal allowance

Income falling under the heading of disregarded income is only taxable to the extent that tax has been deducted at source from it or is treated as deducted or paid, or where the income carries a tax credit.

Other UK source income is taxed in the normal way, but without the deduction of the personal allowance.

For example, UK dividends carry a 10% tax credit and there would be no further liability on the dividend income under this treatment. Bank interest is usually paid net of tax however a non-UK resident can apply to receive UK bank interest gross, and if received gross, there would be no tax liability on this income.

You can choose whichever method produces the lowest tax liability on a year-by-year basis. HM Revenue and Customs (HMRC) will calculate the most beneficial method for you.

The disregarded income treatment is usually more beneficial where the individual has significant amounts of UK savings and investment income compared to other income, such as annuity income, government service pension or rental income.

"The hardest thing to understand in the world is the income tax."

ALBERT EINSTEIN

Income Tax

Income tax can have implications for investment income – you need to make sure that your assets are structured in the best way to limit income and capital gains tax on your investments.

We need to talk.
www.blevinsfranks.com

CHAPTER 5

Spanish Capital Gains Tax

5.1 Taxation of capital gains: summary

Taxation of Spanish capital gains is complex, but the flow charts which follow should help in understanding the rules.

Gains on any transfer or sale of assets are taxable at fixed rates of 21%, 25% and 27% as part of 'savings' income. These rates apply to both residents and non-residents alike. The first €6,000 of total investment income and gains is taxed at 21%, the next €18,000 at 25% and any amount over €24,000 is taxed at 27%.

Gains on assets held for less than 12 months are treated as income and are added to other earned income (eg, pension income, rental income, employment or self-employment income) and are taxed at the scale rates of tax, not at the 'savings' rates.

5.2 Scope of Spanish capital gains tax (CGT)

The following three conditions must all be fulfilled for a chargeable gain or an allowable loss to arise in Spain:

1. There must be a change in the value of taxpayer's wealth; *and*
2. There must be a change in the composition of the taxpayer's wealth; *and*
3. The profit realised must not have the quality of income.

Eg, if a property is sold in 2013 for the same price as it was bought for in 2001, although there is a change in the composition of the taxpayer's wealth, there is no change in the value, so no chargeable gain arises.

Eg, salary received by a taxpayer will fulfil the first two conditions, but as this is income in nature, it cannot be a chargeable gain for Spanish tax purposes.

There are three main types of chargeable gain:

1. Transfer of an asset for consideration.
2. Transfer of an asset for no consideration (gifts, inheritances).
3. Additions to taxpayer's wealth (eg, prize winnings that are not otherwise exempt).

Transfer of an asset for consideration

This is the most common type of transfer. The CGT rules are detailed from Section 5.3 onwards.

Inheritances and gifts

INHERITANCES

An inheritance received by a beneficiary is liable to succession tax not capital gains tax in the hands of the beneficiary.

There is specifically no capital gains tax on death.

GIFTS

Gifts received by an individual are chargeable to gift tax (not CGT) in the hands of the recipient, BUT, in the hands of the donor if a chargeable gain arises it is taxable according to the normal capital gains tax rules.

A lifetime gift cannot ever give rise to an allowable loss.

The transfer value is the same as that used to calculate the gift tax.

Eg, Mr A decides to gift a property to his son. As this produces a change in the value and composition of his wealth which is not income, a chargeable gain arises.

The gift in the hands of the son is liable to gift tax.

5.3 Exemptions

- Assets divided on divorce or the dissolution of a community marriage regime.
- Gifts to certain charities and non-profit-making entities.
- Main home for the over 65s.
- Certain shareholdings.

5.4 Assets acquired on or before 31 December 1994

There are certain deductions available for assets acquired on or before 31 December 1994. Details of how these deductions are calculated for each type of asset are given below. The rules are different for Spanish real estate, any quoted shares and all other assets.

Where an asset was acquired before this date, the gain needs to be time-apportioned into the gain arising before 20 January 2006 and the gain accruing from that date. The deductions are only available on the portion of the gain accruing before 20 January 2006.

5.5 Capital gains on real estate

In addition to any mainstream CGT on the sale of Spanish property there is also a local tax in urban areas commonly known as the *plusvalía* (which literally translates as 'capital gain') on the growth in the value of urban land excluding the buildings. Any *plusvalía* tax paid is allowed as a cost of disposal in calculating the mainstream CGT.

Real Estate Gains

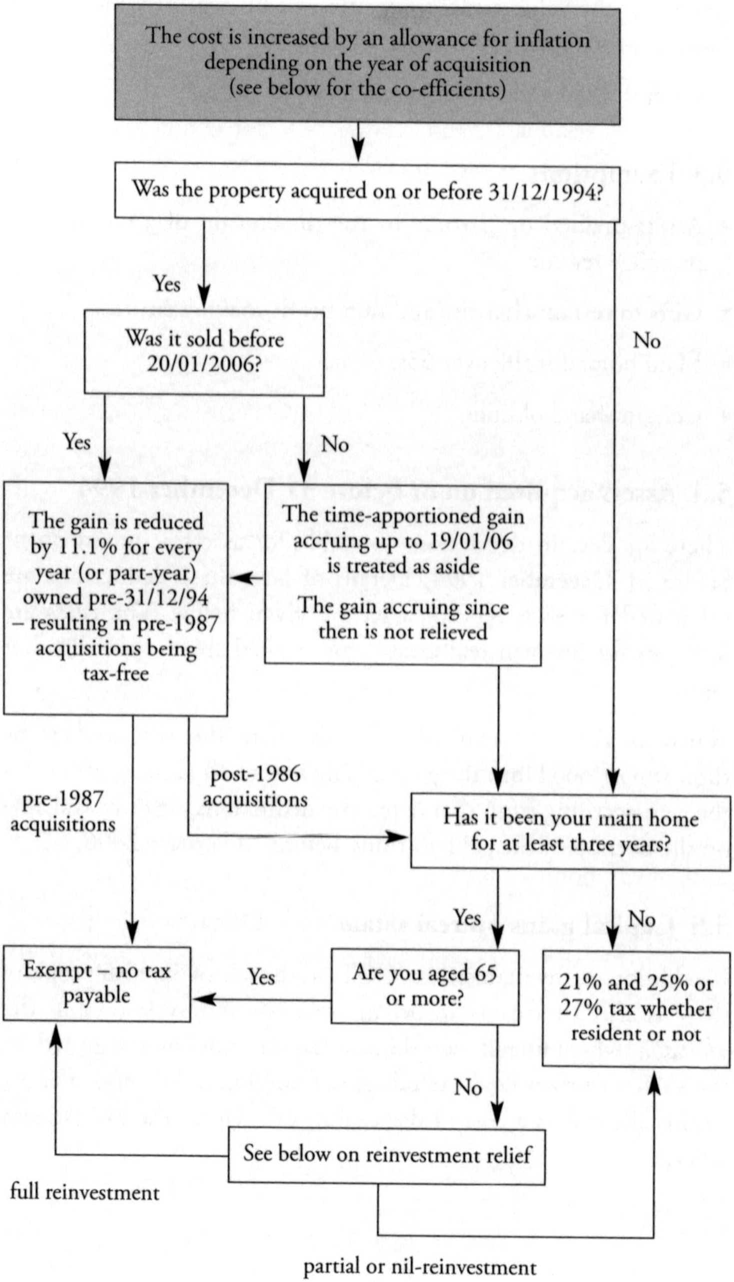

Sale of main home and reinvestment into new main home

UNDER 65 YEARS OF AGE

Tax relief on the disposal of a main residence is only available to Spanish tax residents. Non-residents cannot obtain the relief on Spanish real estate, as they are not living there. Even those who are living in Spain and wish to reinvest the proceeds into a new main home should note that they are unlikely to receive this relief if they are not registered as a resident and paying taxes in Spain. Therefore, they should consider if they should be completing Spanish tax returns, even if these are not required in Spain.

To qualify for the relief you must:

- Have lived in the house for a continuous period of at least three years (or, if less, you had to sell because of a change of job, marriage, separation or the death of a partner) from the date the property was bought, or construction was completed; *and*
- You must buy the new main residence within a period of four years, starting two years before the sale.

A property will qualify as the main home if it is the main home at the date of sale OR has been the main home up to anytime during the two years prior to the date of sale. So, there is a two-year window in which to sell the property after an individual has moved out. Provided the property is sold during this period, it will still qualify as the main home.

The tax relief is based on the proportion of the total sales proceeds reinvested in the new home. If the new home costs more than the sale price of the old home, then the full gain is exempt from tax.

If only half of the sale proceeds are reinvested, then only half of the gain is exempt from tax and the other half is taxable in the year of sale.

However, if the property being sold has a mortgage on it, then it is the net sale proceeds (after deducting the mortgage) which need to be fully reinvested to escape capital gains tax in Spain.

In order for the relief to apply, the taxpayer must declare the gain on their Spanish tax return together with their intention to reinvest the proceeds into a new main home. If the required declarations are not made, the relief is likely to be denied by the Spanish tax authorities. This follows a case heard in Galicia in 2007.

The main residence does not need to be in Spain to qualify for the relief and, similarly, the new reinvestment home does not need to be located in Spain.

So, if you move to Spain and sell your previous main home (in your former country of residence, eg, UK) after you become Spanish resident, then you may be able to avoid Spanish capital gains tax under these provisions. For the relief to apply, you would need to purchase your Spanish home either within the two years preceding the date of sale of the UK property or within two years following the date of sale of the UK property. In addition, the purchase price of the Spanish property should be at least equal to the sale proceeds of your UK property (net of any outstanding mortgage) for full relief to apply. If the Spanish property costs less than the amount you receive from the sale of your UK property, you will be charged tax in Spain on a proportion of the gain equivalent to the sale proceeds not reinvested.

Similarly, if you sell your main home in Spain and move to another country, the relief can still apply to the former Spanish home even though the new home is not located in Spain. The purchase of the new home must take place within the relevant time limits and the full sale proceeds must be reinvested. If the family members move overseas to the new home, but the taxpayer remains in Spain, the relief will not apply as the property has not become the taxpayer's new main home.

If you buy the new home with the aid of a mortgage before sale of the old home then applying the sale proceeds against the mortgage within the two-year time-limit counts as reinvestment (at purchase). If you are able to buy the new home from your own resources unaided then you cannot have the relief as in no sense can you be said to have directly re-invested. Also, if you cannot sell the old home within two years and thus effectively cannot be seen as having re-invested the proceeds within two years before sale you will lose the relief.

There are special rules for new or re-build properties which extend the reinvestment period forward by another two years (four years from the commencement of construction), but that does not assist the prior reinvestment position.

65 YEARS OR OVER

If the property sold qualifies as your habitual main residence, meaning you have lived in the property for over three years, and are Spanish tax-resident and registered as such, the gain is tax-free and there is no need to buy a new property.

Again, the property will qualify as the main home if it is the main home at the date of sale OR has been the main home up to anytime during the two years prior to the date of sale. So, there is a two-year window in which to sell the property after an individual has moved out. Provided the property is sold during this period, it will still qualify as the main home.

Capital expenditure

Capital expenditure incurred on improving or enhancing the property is allowable as a deduction when calculating the net gain. Repairs and maintenance expenses are not allowable for CGT purposes but are allowable against rental income.

Inflation co-efficients

The purchase price of the property is indexed upwards, to account for inflation adjustments, using the following table for 2013:

Year of purchase	Co-efficient
1995 and prior years	1.3167
1995	1.3911
1996	1.3435
1997	1.3167
1998	1.2912
1999	1.2680
2000	1.2436
2001	1.2192
2002	1.1952
2003	1.1719
2004	1.1489
2005	1.1263
2006	1.1042
2007	1.0826
2008	1.0614
2009	1.0406
2010	1.0303
2011	1.0201
2012	1.0100
2013	1.0000

Real estate acquired on or before 31 December 1994

The taxable gain on real estate acquired on or before 31 December 1994 is reduced by 11.1% for every year (or part thereof) that it was owned before that date. However, the reduction does not apply to the portion of the gain accruing after 20 January 2006.

Where a property was acquired before 31 December 2004, the gain needs to be split into two parts: the portion of the gain arising before 20 January 2006, and the portion of the gain accruing from that date. The deductions are only available on the portion of the gain accruing before 20 January 2006.

5.6 Capital gains on quoted shares

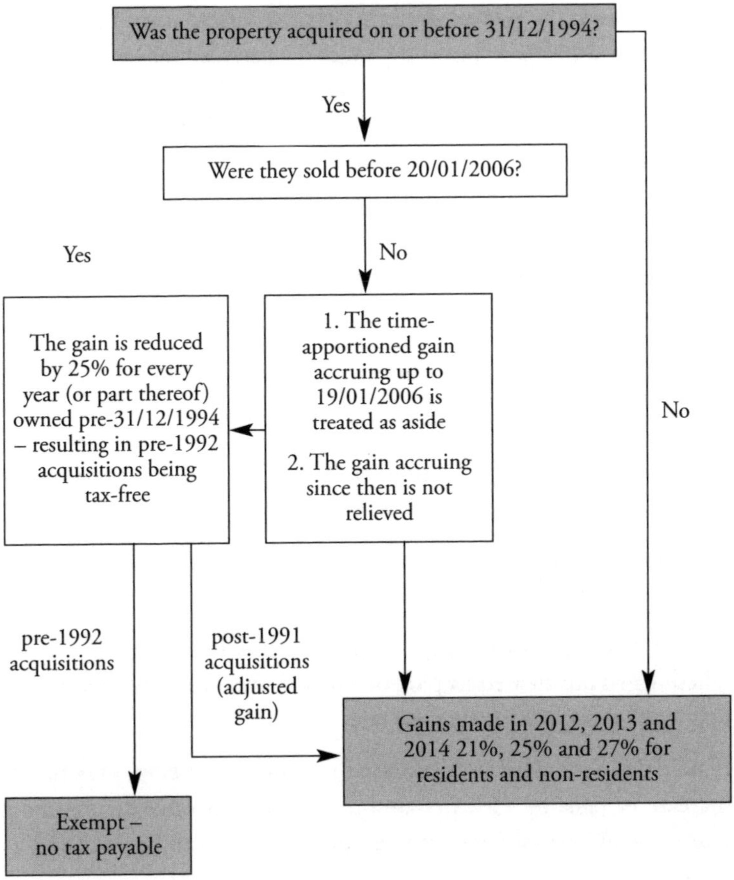

In a similar manner to the treatment of gains on property, the part of the gain on quoted shares accruing prior to 20 January 2006 is reduced by 25% per year for every year or part thereof of ownership prior to 31 December 1994, with the effect that shares acquired before 1992 are exempt on that part of the gain. The part of the gain accruing after 20 January 2006 is simply taxable in full, as is the whole gain on shares acquired after 31 December 1994, since no distinction is made either side of 20 January 2006, in that case.

5.7 Other capital gains

In a similar manner to the treatment of gains on property and quoted shares, the part of the gain on other assets accruing prior to 20 January 2006 is reduced by 14.28%, per year for every year or part year of ownership prior to 31 December 1994, with the effect that assets acquired before 1989 are exempt on that part of the gain. If the asset was sold on or after 20th January 2006, then the part of the gain accruing from that date is simply taxable in full, as is the whole gain on any such asset acquired after 31 December 1994, since no distinction is made either side of 20 January 2006, in that case.

The rates of tax are:

- 2006 – 15% (or 35% for non-residents, subject to tax treaty);
- 1 January 2007 to 31 December 2009 – 18%;
- 1 January 2010 to 31 December 2011 – 19% on the first €6,000 and 21% thereafter;
- From 1 January 2012 onwards – 21% on the first €6,000, 25% on the next €18,000 and 27% thereafter.

Special rules apply to collective investments held in a tax haven. These gains are treated as part of normal income and liable to the taxpayer's marginal income tax rate.

If no data is provided, it is assumed that investments in a tax haven increase in value by 15% per annum and tax is payable each year on that basis. When sold, the 15% increase in value can be deducted as a cost.

Roll-over of gains on *IIC* investments

IIC (Instituciones de Inversión Colectiva) are Spanish approved public access collective investment vehicles that invest in unquoted investment funds. Capital gains made on the sale or reimbursement of the investment funds can be rolled-over, deferring the gain until eventual sale, into new *IIC* holdings, so long as the investor at no time has access to the released funds/money. (Note: dividends from *IIC* investments do not attract the €1,500 income exemption.)

The roll-over relief is not available on transactions that involve other sorts of vehicle (such as a life assurance policy wrapper) even though the underlying investment fund(s) might be of the same sort. The tax regime applying to the old and new investments has to be the same – they cannot be different vehicles. So, it has to be *IIC* proceeds into another *IIC*.

Qualifying investments (or qualifying compartmentalised sub-funds) must have more than 500 shareholders/investors, and to get the tax relief the taxpayer can hold no more than 5% in the past 12 months.

The relief is not available where the *IIC* invests in quoted investment funds.

5.8 Capital gains on Spanish assets made by non-Spanish residents

For sales of Spanish property where the vendor is non-resident, purchasers must withhold 3% of the purchase price (not the gain) and pay it over to the Spanish tax authorities as an advance payment on behalf of the vendor of CGT. If it is not paid, the purchaser can be fined and the unpaid tax becomes a charge over the property itself. The vendor must file a Spanish tax return on the transaction within three months of the sale before any repayment can be made. It should also be noted that transfer tax and VAT are payable, and any sale may also be liable to land appreciation tax (which is a local tax based on increase in land value, known as *plusvalía*).

CGT might not be due if the sale is not of the property, but of a company which owns the property, and that company is resident in a tax treaty country (eg, UK). Gibraltarian companies which own Spanish property would be liable for CGT, if these companies are sold.

EU residents are generally exempt from CGT in Spain on non-real estate gains with the exception of:

- Capital gains obtained through tax havens.
- Capital gains on shares in companies the main assets of which are Spanish real estate. A gain by a company incorporated in a country with a suitable double tax treaty may be tax-free. However, a Gibraltarian company would be liable for tax.
- Capital gains where the non-resident owned at least 25% at any time within the previous 12 months.

However, any applicable double tax treaty needs to be reviewed as this might reduce or eliminate the tax.

The Spanish authorities reserve the right to review the disposal price if they believe it to be understated by more than 30% or €15,025.30. The difference between the tax valuation and the declared price will be considered a gift. The vendor will be assessed to additional CGT on the difference, and the purchaser will be liable for gift tax.

5.9 Capital losses

Losses may be offset against other capital gains of the current year. Net losses can be carried forward against gains for the next four years.

5.10 The UK/Spain double tax treaty and capital gains

Article 13 of the UK/Spain double tax treaty grants exclusive rights over capital gains, save those made on the disposal of real estate, to the country of residence of the person making the gain. Consequently, in the case of a Spanish resident selling any UK assets other than UK-sited property, and including substantial shareholdings in UK companies, such capital gains can only be taxed in Spain, under the rules described above. However, beware the UK's five-year rule for temporary non-residents (see **Chapter 16**).

5.11 Temporary Spanish residence and CGT

The combined effect of the UK/Spain double tax treaty's tax allocation rules and the Spanish CGT rules make it possible for any CGT liability in the UK and Spain to be substantially reduced when the assets have been held since before 31 December 1994, and the individual making the disposal is Spanish resident at the time of sale.

It is, of course, essential to properly arrange the transfer of residence to Spain so that the previously UK resident individual has clearly become Spanish tax resident and this cannot be challenged by the UK authorities. Special care must be taken because of the fact that the Spanish and UK tax years start at different dates (1 January and 6 April respectively).

On the other hand, such temporary migration to Spain requires careful planning so as to avoid substantial Spanish income tax liabilities, as all Spanish residents are exposed to tax in Spain on their worldwide income.

CHAPTER 6

Spanish Wealth Tax

6.1 Introduction

On December 2008, the Spanish Government approved a measure to apply a 100% tax credit against an individual's wealth tax liability from 1 January 2008, and to remove the requirement to file a wealth tax return. This measure effectively abolished wealth tax from that date.

However, in September 2011, wealth tax was reinstated as a temporary measure, to apply for 2011 and 2012 (ie, assets held at 31 December 2011 and 2012 – tax payable in 2012 and 2013). This has been extended for a further year, to wealth held as at 31 December 2013 (tax payable in 2014).

Spanish wealth tax is payable by non-residents and residents based on assets held at 31 December each year. Spanish residents are liable on their worldwide assets, and non-residents are liable to wealth tax on their Spanish assets only.

6.2 Rates and allowances

Allowances for 2011 to 2013:

	Residents	**Non-Residents**
Assets	Worldwide	Spanish assets only
Deductions		
Individual	€700,000	€700,000
Own Home	€300,000	None
Example of a Married Couple		
2 x Individual	€1,400,000	€1,400,000
2 x Home	€600,000	None
Total Deductions Available	€2,000,000	€1,400,000

Example assumes assets are held jointly and the main home is in joint names and valued at over €600,000.

It is possible for the Autonomous Regions to vary the allowances and rates, and in the past, certain regions have increased the state allowances generally and also for those with disabilities.

2013 State Tax Rates

From €	To €	Band €	Tax Rate %	Tax Band €	Cumulative Tax €
Nil	167,129	167,129	0.2	334	334
167,129	334,253	167,124	0.3	501	836
334,253	668,500	334,247	0.5	1,671	2,507
668,500	1,337,000	668,500	0.9	6,017	8,523
1,337,000	2,673,999	1,336,999	1.3	17,381	25,904
2,673,999	5,347,998	2,673,999	1.7	45,458	71,362
5,347,998	10,695,996	5,347,998	2.1	112,308	183,670
Over	10,695,996	–	2.5	–	–

2013 Rates in Andalucía

From €	To €	Band €	Tax Rate %	Tax Band €	Cumulative Tax €
Nil	167,129	167,129	0.24	401	401
167,129	334,253	167,124	0.36	602	1,003
334,253	668,499	334,247	0.61	2,038	3,041
668,499	1,336,999	668,500	1.09	7,287	10,328
1,336,999	2,673,999	1,336,999	1.57	20,990	31,318
2,673,999	5,347,998	2,673,999	2.06	50,084	81,402
5,347,998	10,695,996	5,347,998	2.54	135,839	217,241
Over	10,695,996	–	3.03	–	–

6.3 Exempt assets

The following assets are exempt from wealth tax:

- Owner managed small businesses.
- Family companies meeting certain conditions (see below).
- Shares in property investment companies are not exempt unless the company carries on a commercial activity (see below).
- Household contents (excluding jewels, fur coats, vehicles, boats, art, and antiques).
- Pension rights (but not the capitalised value of an annuity).
- Intellectual property rights in the author's ownership.
- Business assets. For the assets to qualify as business assets, the activity must be the taxpayer's main source of income (ie, the income from the business must constitute at least 50% of his taxable income), and the activity must be carried out by the taxpayer on his own account and on a habitual basis.

Rental/property investment businesses

Where a rental/property development business is carried out, the following conditions must be fulfilled for the activity to qualify as a commercial activity:

- There must be a building or premises used **exclusively** for the management of the business activity. Part of a building can qualify provided the part used is separate from any other activity and is used exclusively for the management of the property business. A shared office will not qualify as separate premises.
- There must be at least one member of staff employed on a full-time contract. This could be a spouse but the spouse would need to be registered as an employee for social security in Spain and social security contributions would need to be deducted from their salary each month.

Provided these conditions are fulfilled, the properties used in a rental/development business can be exempt from wealth tax in Spain.

Shareholdings

Shareholdings are exempt from wealth tax provided:

- The company is a trading company; *and*
- You own at least 5% of the share capital (or at least 20% including shareholdings belonging to a spouse or other family members); *and*
- You carry out managerial duties for the company (including President, MD, Director of a specific department, general manager, etc, provided you have a say in the decisions taken regarding the company); *and*
- You derive a salary for such activities which is at least 50% of your total net earnings.

Deductibility of loans

A loan will be deductible in arriving at an individual's net wealth tax liability as long as the loan was not used to buy or invest in assets exempt from Spanish wealth tax. Exempt assets are listed in the three preceding sections above.

In the case of non-residents with non-Spanish loans secured on Spanish property, deduction should be possible on a similar basis to that ruled on for succession tax.

6.4 Limits to wealth tax

Cumulative wealth and income taxes cannot exceed 60% of:

- The 'general and savings taxable income bases' of residents (but still excluding from savings income any gains on assets held for more than one year, and the associated 21%/25%/27% tax).
- Subject to paying a minimum of 20% of the full wealth tax calculation on income producing assets.
- Wealth tax on non-income producing assets cannot be reduced.

There is no limit for non-residents.

Income and non-income producing assets

A PCP (whether held in an individual's own name or in trust) is an income producing asset for Spanish wealth tax purposes (*DGT* 21-6-96). This is the case even if no withdrawals have been made so no taxable income has arisen. However, there are certain types of such bonds that are not subject to wealth tax in certain circumstances.

Where a loan has been made to a trust by a Spanish resident individual, the loan will be ignored for wealth tax purposes and the underlying asset(s) in the trust will be charged to wealth tax as if held personally by the individual. This is because the trust is treated as transparent for Spanish tax purposes. This is a favourable position if the underlying asset is a PCP, since this is an income producing asset and the wealth tax on it can be restricted, whereas an interest-free loan would be a non-income producing asset and the wealth tax on it could not be restricted.

You exclude from the calculation any wealth tax on assets that by their nature or intended purpose are not capable of generating taxable income (eg, antiques and works of art, jewellery, private cars, aircraft, boats, unbuilt land, **interest-free loans**, main home, or other property that is not rented out). **The wealth tax on such assets therefore remains payable in full and unreduced.** It is calculated as a straight proportion of the overall tax (so, not the top slice). There is an example of how this works in the examples below.

An interest-free loan (where it is properly documented as such) will be treated as a non-income producing asset for the purpose of this restriction. In this case, full wealth tax would be payable in relation to the loan.

EXAMPLES

From 1 January 2007, the income to be taken into account is the combined general and savings taxable income bases with deduction of the same basic 'minimo' allowance (even though for other purposes all allowances now go after the 'taxable' *(imponible)* base in arriving at the 'net' *(liquidable)* base). The idea is to keep the wealth tax restriction based on the same parameters as before the 2007 reforms.

So, although from 1 January 2007, **all** capital gains are included in savings income, those on assets held for more than one year are still excluded from the savings income taxable base for the purposes of calculating the wealth tax ceiling.

EXAMPLE

For instance, the taxpayer has the following income/capital gains and losses:

Business income	€100,000
Interest	€25,000
Capital gain (more than one year)	€30,000
Capital loss (less than one year)	€20,000

Under 2007 rules

General income tax base	€100,000
Savings income tax base €25,000 + €30,000 - €20,000	= €35,000
Total tax base	€135,000
Capital gain (> one year)	€30,000

Tax base for combined taxation wealth tax and personal income tax (60%):

€(135,000 - 30,000) x 60% = **€105,000** (see below item **C**)

Taxable general base Personal Income Tax: €(100,000 - 5,151[1]) = €94,849

Taxable saving base Personal Income Tax: €35,000

Personal Income Tax (PIT) liability = €41,300 (estimated amount)

[1] Personal reduction established by law *(minimo)*

a. The PIT liability which must be taken into account for the 60% limitation will be €41,300 - 6,300 (tax derived from capital gain > one year = 21% x 30,000) = **€35,000.**

b. The wealth tax liability might be, for example, **€90,000** (this amount is hypothetical and based all on income producing assets).

(A) + (B) = **€125,000**

(C) Limitation (60% of PIT base) = **€105,000**

Excess = (A) + (B) - (C) = **€20,000**

Therefore, the wealth tax due would be: €90,000 - 20,000* = €70,000

*The reduction is less than the 80% (ie, the residual wealth tax is not less than 20%) of the full amount – so the reduction stands.

But what if the wealth tax had included €20,000 tax on non-income producing wealth – ie, 2/9 of the assets were non-income producing (and the income tax figures were the same)?

(A) + (B) = €35,000 + (90,000 - 20,000) = **€105,000**

Excess = **€0**

The wealth tax would be: €(70,000) + €20,000 = €90,000

Limits to wealth tax where a joint income tax return is filed

Where a married couple have opted to file a joint income tax return, in order to calculate the wealth tax limitation, you add together the total income tax due and each individual wealth tax calculation. If the 60% limit is exceeded, the reduction in wealth tax is pro-rated between the individuals in proportion to the amount of their taxable wealth.

EXAMPLE

A married couple's total general and savings taxable income base (joint tax return) is €14,000 (after the relevant income exclusions and basic *'minimo'* allowances), and their income tax liability is €2,000. The husband's individual wealth tax liability is €6,000 and the wife's is €9,000. Total income taxes and wealth taxes are therefore €17,000.

Combined income and wealth taxes cannot exceed 60% of taxable income, ie, 60% of €14,000 = €8,400. Excess taxes are therefore €8,600. The 20% minimum does not apply.

The couple's individual wealth taxes are reduced by:

Husband €6,000/€15,000 x €8,600 = €3,440

Wife €9,000/€15,000 x €8,600 = €5,160

Final wealth taxes are:

Husband €6,000 - €3,440 = €2,560

Wife €9,000 - €5,160 = €3,840

In certain circumstances, therefore, it may not be favourable for wealth tax purposes for a married couple to file a joint income tax return. This could be the case where one spouse has little or no taxable income but a substantial asset base.

6.5 Property value

Property is valued at the highest of:

- The officially registered *valor catastral*; *and*
- The value taken into account for any other tax purposes; *and*
- The price in the purchase agreement.

Liabilities in general reduce taxable wealth, but not where it is a loan used to buy an asset that is specifically exempt or covered by exemptions. So where a mortgage is for the purchase of the main home the value of which for wealth tax is covered by the main home exemption, no deduction is available for that mortgage.

For a non-resident, only Spanish liabilities would be taken into account but there is no exemption to consider. A Spanish mortgage would have to be attached to a Spanish property to be deductible for wealth tax purposes.

Valuing a usufruct (usufr)/freehold interest

Wealth tax is payable by both the holder of the usufruct and the holder of the freehold interest *(nuda propriedad).*

The value of the usufruct is calculated as 70% of the value of the property where the usufruct holder is less than 20. As the age of the usufruct holder increases, the value of the usufruct reduces by 1% for each year over 20 to a minimum of 10% of the value of the property.

Age usufr holder	Value usufr %	Age usufr holder	Value usufr %	Age usufr holder	Value usufr %	Age usufr holder	Value usufr %
19	70	35	54	51	38	67	22
20	69	36	53	52	37	68	21
21	68	37	52	53	36	69	20
22	67	38	51	54	35	70	19
23	66	39	50	55	34	71	18
24	65	40	49	56	33	72	17
25	64	41	48	57	32	73	16
26	63	42	47	58	31	74	15
27	62	43	46	59	30	75	14
28	61	44	45	60	29	76	13
29	60	45	44	61	28	77	12
30	59	46	43	62	27	78	11
31	58	47	42	63	26	79	10
32	57	48	41	64	25	–	–
33	56	49	40	65	24	–	–
34	55	50	39	66	23	–	–

The value of the freehold interest is calculated as the difference between the value of the property and the value of the usufruct interest as calculated above.

6.6 Purchased annuities

Wealth tax is payable on whole of life and temporary purchased annuities. The assessable value is determined by capitalising the annuity using the official Bank of Spain base interest rate for the year (eg, 4% for 2013).

For **Whole of Life annuities**, you then apply to the capitalised value the relevant percentage using the usufruct tables above, based on the age of the person receiving the annuity. Your age is taken at 31 December each year, and so as you get older, the value to be included in your wealth tax return will decrease.

EXAMPLE

John is 65 at 31 December 2013. He receives a Whole of Life annuity of €25,000 per year. **For 2013**, his wealth tax liability is:

- The income is capitalised using the official rate of interest in 2013 (4%): €25,000 x 100/4 = **€625,000.**
- The relevant percentage is taken from the usufruct tables for an individual who is 65: **24%.**
- The above percentage is applied to the capitalised value: €625,000 x 24% = **€150,000**. This is the value that should be entered on John's wealth tax return for 2013.
- The amount of wealth tax payable will of course depend on his marginal rate of wealth tax. However, assuming this is 0.5%, the wealth tax in the first year would be €750.

For **2014**, assuming the official rate of interest has not changed, the value for wealth tax will be determined by multiplying the capitalised value by 23%, ie, €625,000 x 23% = **€143,750.** At a rate of 0.5%, the wealth tax would be €719.

Note, the age related percentage reduces by 1% each year as the individual's age increases, so the amount subject to wealth tax will reduce by €6,250 each year (assuming that the official rate of interest does not change). This will result in a reduction in the wealth tax liability of €31.25 per year, where the rate chargeable is 0.5%. Wealth tax liability will also change depending on the official exchange rate in force in the year concerned.

For **Temporary annuities**, you multiply the capitalised value by 2% for each year the policy has left to run, up to a maximum of 70%.

ILLUSTRATION

John purchases a temporary annuity of €25,000 per year over a 10-year period.

1. The income is capitalised using the official rate of interest, say 4%: €25,000 x 100/4 = **€625,000**. Note this example assumes a constant rate of interest.

2. The relevant percentage is taken depending on the number of years the policy has left (up to a maximum of 70%).

   Year 1 – 2% x 10 years = 20%

   Year 2 – 2% x 9 years = 18%

   Year 3 – 2% x 8 years = 16%

   Year 4 – 2% x 7 years = 14%

   Year 5 – 2% x 6 year = 12% and so on

3. The above percentage is applied to the capitalised value:

   Year 1 – €625,000 x 20% = €125,000

   Year 2 – €625,000 x 18% = €112,500

   Year 3 – €625,000 x 16% = €100,000

   Year 4 – €625,000 x 14% = €87,500

   Year 5 – €625,000 x 12% = €75,000 and so on

4. The amount of wealth tax payable will of course depend on John's marginal rate of wealth tax. However, assuming this is 0.5%, the wealth tax in the first five years would be as follows:

   Year 1 – €125,000 x 0.5% = €625

   Year 2 – €112,500 x 0.5% = €562.5

   Year 3 – €100,000 x 0.5% = €500

   Year 4 – €87,500 x 0.5% = €437.50

   Year 5 – €75,000 x 0.5% = €375

Note, the relevant percentage reduces by 2% each year as the annuity period expires, so the amount subject to wealth tax will reduce by €12,500 each year. This will result in a reduction in the wealth tax liability by €62.50 per year, where the rate chargeable is 0.5%. Wealth tax liability will also change depending on the official exchange rate in force in the year concerned.

6.7 Bank balances

Bank balances are valued at the higher of the closing balance on 31 December or the average balance during the 4th quarter. This information is automatically given by the bank.

6.8 Trust assets

As trusts are now treated as transparent in Spain, Spanish resident settlors of trusts should declare their assets as their own if they are still alive.

Beneficiaries of life interest trusts where the settlor has died should declare the full value of the trust assets over which they have a life interest for wealth tax in Spain, as these assets are treated as theirs.

The position is not as clear for beneficiaries of discretionary trusts, and we recommend that advice is sought.

6.9 Returns and payments

The wealth tax form must be completed and tax is payable between 2 May and 1 July for the previous 31 December. Husband and wife make separate returns reflecting their shares of any joint assets and liabilities in addition to any personal items.

CHAPTER 7

Spanish Reporting Obligations

7.1 Introduction

From 2013, there has been a requirement for residents of Spain to report certain classes of offshore (non-Spanish) assets to the Spanish Government held as at 31 December the previous year.

This is not a tax, and the information given is not part of any tax return, simply a requirement to report information about assets, although it is reasonable to assume that the form will be compared with any wealth tax return due.

7.2 Who, what, when and how?

Who is liable to make a report?

Only Spanish residents are required to make a report of their non-Spanish assets every year.

The report is required where the individual is the owner, beneficiary or authorised signatory, including those with a power of attorney. This therefore includes assets held by a trust or fiduciary.

What is reportable?

Spanish residents are required to make a report declaring their share of any foreign assets falling within various asset classes each year. The report is required where the assets (over which they have the rights above) in that class exceed €50,000, even if the individual's personal share is valued at less than €50,000.

Once the report is made, if no more assets are acquired, and the existing assets with that class of assets increase in value by less than €20,000 and are retained, no further report is required in subsequent years. If other assets are acquired, or the existing assets in that class increase in value by more than €20,000 (even if their share of those assets increase by less than this amount) or are sold, a report is required for that year.

By when?

The due date for reporting was 31 March in respect of assets held as at 31 December the previous year (so by 31 March 2014 for assets held as at 31 December 2013).

For 2013 only (reporting assets held as at 31 December 2012), the due date for reporting was 30 April 2013.

How?

The reporting form is *Modelo 720*, and this was published following Minsterial Order *Orden HAP/72/2013* of 30 January.

The form is to be completed on-line, and so is not in standard format. Rather, there are a series of drop down menus and the user is guided through each stage, and in many cases, simply completes a number or letter in the relevant field.

We would recommend that our clients always seek the assistance of an accountant when completing the form.

7.3 What are the classes of assets?

Strictly, there are three classes of assets:

1. Accounts held with financial institutions.

2. All types of immovable property, and rights over such property.

3. Shares and securities and other assets, which include:

- Life insurance policies.
- Temporary or lifetime income generated from the lending of money, rights or other assets (including immovable) to foreign entities.

7.4 What needs to be reported?

Life Insurance Policies

- Full details of the insurance company, including name, address and overseas ID (or tax registration) number.
- Date the policy was taken out.
- Surrender value as at 31 December.
- Ownership %.
- If redeemed/surrendered during the year, the date of surrender and the value at that date.

Bank Accounts

- Details of the bank, address, and overseas ID (or tax registration) number.
- Sort code, account number and type of account (eg, current, savings).
- Date the account was opened.
- Balance at 31 December **and** average balance over last quarter.
- Ownership %.
- If closed during the year, the date of closure and the value at that date.

Shares/Securities/Equities/Unit Trusts, etc

- Details of the company/investment house.
- Number of shares/units.
- ISIN number where available.

- Date of acquisition.
- Value at 31 December.
- Ownership %.
- If disposed of during the year, the date of disposal and the value at that date.

Immovable Property

- Full address and post code.
- Date of acquisition and cost of acquisition, including any taxes paid at acquisition.
- Date of disposal, where relevant, and value at that date.
- Urban or rustic.
- Ownership %.
- Value at 31 December where there are rights of usufruct or *nuda propiedad*, or other property rights.

7.5 Q&As from *Hacienda* website

The *Hacienda* have published a list of Q&As on their website in relation to the new reporting rules, giving guidance on how to complete the new reporting Form 720.

The Q&As can be found at:
http://www.agenciatributaria.es

We have not been able to locate a Q&A section for *Modelo 720* for 2014 reporting of assets held in 2013, so we would advise that the 2012 Q&As are used.

Below are the main points of interest. Note that the ordering does not follow that on the website:

Q Where there are joint or multiple owners of an overseas bank account and the total balance is more than €50,000, does the account need to be reported?

A The obligation to report exists where the balance of the account is over €50,000 regardless of how many account holders there are. Each account holder should declare the total balance (not pro-rated), indicating the percentage they hold.

Q Does the form have to be completed when an overseas property has a value of more than €50,000 as at 31 December, but it is owned by several people?

A As the value of the property is over €50,000, this will trigger the obligation to report, irrespective of how many owners there are. The total value should be declared on the declaration, indicating the individual's percentage shareholding.

The same principle will apply to all assets and rights, etc, where they are owned by several individuals.

Q If, for example, an individual ceases to be the account holder in June of the relevant year, is there an obligation to make the declaration?

A The obligation to file the form exists if the balance in the account on that date is over €50,000 (when combined with other overseas accounts). Full details will need to be given about the account, the credit entity it was held with, and the balance on the date the individual ceased to be the account holder.

Q Where an overseas account is over €50,000 and a declaration is duly filed, and in a following year the balance is below €50,000 and the individual closes the account, is there are requirement to report?

A Yes, because the account has been closed in the relevant year.

Q Do overseas pension plans need to be declared?

A There is no obligation to provide information on pension plans situated overseas provided that they do not give rise to a temporary or lifetime annuity.

Where there is a Temporary or Lifetime annuity, the capitalised value, as calculated for wealth tax purposes, should be declared.

Q Do stock options have to be declared?

A No.

Q If an individual moves overseas once the tax year has started and he has to file an income tax declaration as a resident for this period, does he need to file a reporting form?

A Yes, provided that the conditions for reporting are met.

Q If a Spanish resident has an account with an overseas establishment of a Spanish domiciled entity, does he have to make a report?

A No, provided that the entity domiciled in Spain has duly reported the account in accordance with Royal Decree 1065/2007 of 27 July and provided that the account was able to be declared in accordance with the legislation in the country where it is situated.

Q How do you value shares and rights in a company?

A The valuation should be calculated using the wealth tax rules.

Q If an individual acquired a property as a result of a gift or an inheritance, what is the acquisition value that needs to be declared?

A If the property has been acquired as a result of a gift or an inheritance, the actual acquisition value needs to be declared ie, the real value of the asset at the time of its original acquisition.

Q Is it necessary to declare the values of all shares that have been sold during the year and reinvested as a consequence of transactions involving sales and reinvestments?

A No, provided the proceeds are wholly reinvested in the acquisition of other shares or rights which are in turn declared, you only need to declare the values as at 31 December.

Q Where an account is held in a different currency, what is the exchange rate that should be used? And what if the account was closed during the year?

A The exchange rate should be that in force at 31 December each year and the same rate should be used for valuing the average balance of the last quarter. Where the account was closed during the year, you should use the exchange rate at the date of closure when reporting the account balance.

Q Where an overseas property was bought in a different currency, what exchange rate should be used to determine the acquisition value?

A The acquisition value should be converted into Euros at the exchange rate in force at the 31 December of the tax year corresponding to the relevant report.

Q Do you need to take into account variations in the exchange rate when valuing assets each year?

A Yes. For example, for the 2013 declaration (in relation to assets situated overseas held at 31 December 2012), you apply the exchange rate in force at that date. In 2014 (ie, valuation at 31 December 2013) you apply the exchange rate in force at 31 December 2013. There is no requirement to file a declaration again provided that the owner has not disposed of the asset, and provided the total balance of all overseas accounts in that category has not increased by €20,000 compared to the previous year.

Q What are the penalties for failing to report:

a. One overseas account, two other moveable assets, and three immovable properties?

The minimum penalty for failing to file a Form 720 in this example (where all three categories should have been reported) will be €30,000.

b. Assets in just one category?

Where only one category of information should have been reported, the minimum penalty is €10,000.

Q In relation to assets declared on Form 720, in the event that they relate to undeclared income, can this be regularised?

A Yes. If it the asset was purchased with undeclared income, this can be regularised under the normal regularisation procedures.

Q What is the position with overseas collective investments (UCITS) held by Spanish residents?

A Overseas UCITS which are sold in Spain by Spanish entities or representatives in Spain of overseas institutions operating in Spain under the freedom to provide services Directive do not need to be reported on Form 720. This is because such institutions are obliged to provide information about such investments to the Spanish authorities under article 39.2 of *Reglamento General* approved by Royal Decree 1065/2007 of 27 July.

Q If the taxpayer holds a life insurance policy which is contracted with an overseas insurance company operating in Spain under the Freedom of Services regime, does the taxpayer have an obligation to report the policy on Form 720?

A In relation to such policies where the insurance company is operating in Spain under the Freedom of Services regime, there is no obligation to report such policies by virtue of article 42 ter.3 of the General Regulations approved by Royal Decree

1065/2007 of 27 July, provided that in fulfilment of article 39.3a) of such regulations, the representative of the insurance company provides to the Spanish Authorities the information required under the above regulations.

NOTE

However, our advice is that this is only a Q&A for guidance only, and does not have the force of law. If you choose to rely on this guidance, we would recommend that your accountant (or whoever is responsible for submitting the declaration on your behalf) makes a file note that the decision not to report any such policy is based on the guidance issued by the *Hacienda*, and a copy of the relevant Q&A should be printed out and kept with the note, so that if the position is queried, you have evidence as to why the policy was not reported.

7.6 Official Exchange Rate

This is the link to the Spanish official exchange rates:

http://www.boe.es/boe/dias/2014/01/03/pdfs/BOE-A-2014-112.pdf

7.7 Penalties

There are various penalties:

- Late declaration – €1,500.
- Failure to declare – €5,000 per asset per asset class, with a minimum charge of €10,000.
- The asset will be treated as undeclared extraordinary income in the year, so will be taxable in the hands of the owner. The undeclared income arising from the asset will be deemed to arise in the last tax year that is not statute-barred (in most cases, this will be four years ago). This effectively abolishes the application of the statute of limitations for such undeclared income. The undeclared income will be taxable as general income at the income tax scale rates, which can be over 50%.

- If held to be undeclared extraordinary income, a further penalty of up to 150% of the tax due will be levied – which means that more than the value of the asset can be taken in tax and penalties.

  So, for example if the undeclared asset/income has a value of €100, and the tax due on it is 52%, then the penalty will be 150% of €52. This is €78.

  So the total cost will be:

Income tax	€52
Penalty	€78
Total	**€130**

  Plus late payment interest.

In many cases the amount of the unpaid tax could give rise to the criminal offence of 'tax fraud' (where the tax defrauded exceeds €120,000).

If it can be proved that the funds came from income earned while not Spanish resident, or that the funds came from taxed income, the income tax and income tax penalties will not be levied.

CHAPTER 8

Spanish Inheritance and Gift Tax

8.1 Scope of tax

Spanish inheritance and gift tax (SIGT) is both an inheritance tax and a gift tax. The Spanish term for it is *Impuesto sobre sucesiones y donaciones (ISD).* Unlike in the UK, it is the recipient and not the donor's or the deceased's estate that pays the tax, but it is only due if:

- The recipient of a gift or the beneficiary of an estate is resident in Spain; *or*
- The asset being gifted or inherited is 'property in Spain' (ie, a Spanish asset).

'Property in Spain' constitutes any asset or right located in Spain, or which can be legally exercised or realised there, and includes:

- Real estate situated in Spain; *and*
- Moveable property located in Spain or permanently associated with land and buildings in Spain.

To put it another way, SST is not payable if the assets are outside Spain *and* the recipient is not resident in Spain.

The diagram below demonstrates this:

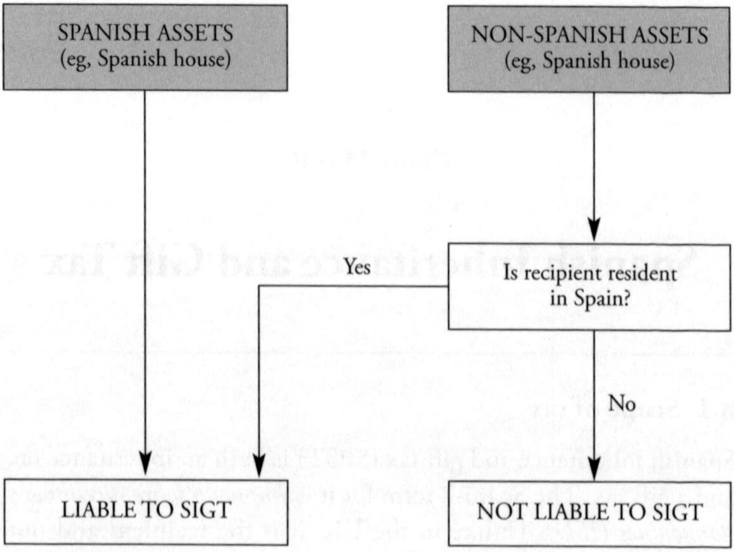

Property-holding companies

The question arises whether the shares in a non-Spanish company that owns Spanish property should themselves be considered Spanish-located by virtue of the underlying real estate asset(s). If not, then the inheritance or receipt by gift of such shares by a non-resident of Spain should not be subject to Spanish succession and gift tax.

However, for this to work, the non-Spanish (eg, UK) company must not be 'effectively managed' in Spain, otherwise this would make it Spanish resident. And if it is not to be managed in Spain then in practice you, the company owner(s), should not be Spanish resident. So, it only works for non-resident companies, run by non-residents, where the shares are inherited by or gifted to non-residents.

Furthermore, any possible Spanish succession and gift tax saving on a gift or inheritance of the shares from you as a non-resident needs to be weighed against the cost of setting up and running the company, and the possibility of your one day becoming resident in Spain. In that case, not only will the succession and gift tax saving disappear, but if the property were to become your main home, holding it in a company will deny you the valuable capital gains exemptions that resident homeowners can have in Spain on a sale of their property.

8.2 Calculation of tax liability

Succession and gift tax in Spain is calculated as follows:

Assets	Amount inherited
+	Other taxable amounts (eg, gifts received from deceased in last three years)
+	General household contents
Less	Exempt assets Charges (eg, mortgages) Debts (eg, income tax of deceased) Allowable expenses (eg, funeral expenses)
=	**Gross Taxable base *'base imponible'***
Less	Allowances (eg, personal, main home, business, etc)
=	**Net taxable base *'base liquidable'***
Then	Apply tax scale rates
Then	Apply multipliers
=	**Gross tax payable**
Less	Reductions (eg, double tax relief, tax paid earlier on gifts now cumulated in the estate)
=	**Total tax payable**

8.3 The state rules

The state (national) rules are set out below. They can be varied regionally. First, beneficiaries are divided into the following four groups depending on the closeness of relationship to the donor or the deceased:

GROUP I	Natural and adopted children and other descendants (eg, grandchildren, great-grandchildren, etc) under 21
GROUP II	Natural and adopted children and other descendants aged 21 and over Parents and other ascendants (eg, grandparents, great-grandparents, etc) Spouses
GROUP III	In-laws and their ascendants/descendants Step-children Brothers and sisters Cousins Nieces and nephews Aunts and uncles
GROUP IV	All others including unmarried partners even if they have registered as a *pareja de hecho* (a *de facto* couple)

State allowances

There are no state allowances for lifetime gifts. The tax-free state allowances for inheritances by members of the different groups are as follows:

GROUPS I and II	€15,956.87
GROUP III	€7,993.46
GROUP IV	nil

In addition, Group I inheritors (children, including adopted children, and grandchildren, etc, in the direct line of descent, under the age of 21) can have an additional deduction of €3,990.72 for each year they are under 21, restricted in total to €47,858.59 per recipient.

Also, where the recipient is physically or mentally disabled, further reductions are available depending on the recognised degree of disability. This reduction is in addition to the other reductions available, and is not dependent on the degree of kinship with the deceased:

Between 33% and 65% disability	€47,858.59
Over 65% disability	€150,253.03

(The rather precise figures found in Spanish tax allowances and rate bands are a hang-over from the days of the *Pesetas*.)

These rates have not changed for some years.

State tax rates and multipliers

The following rates have applied since 1 January 2003. The tax liability is subject to multipliers dependent on:

- The relationship of the recipient to the donor/deceased; *and*
- The wealth of the recipient prior to the transfer ('pre-existent net worth').

The highest effective rate of tax is just under 82% on this basis, though before any multipliers the maximum rate stands at 34%.

It should be noted that, if the heir is worth more than €402,678.11 the tax payable is, in effect, increased by as much as 20%. Note also that the tax cannot be offset against any UK inheritance tax on interspousal inheritance, since there is no UK tax payable except in rare circumstances (just Spanish tax).

However, where a non-resident individual inherits Spanish property, it is only their net Spanish assets (ie, the assets on their Spanish wealth tax return) that are taken into account when calculating their net worth for the multipliers. If, for example, the non-resident inheritors are the children of the deceased and they do not have any assets in Spain, the multipliers will not be an issue.

Inheritance and Gift Tax Rates – 2013 and 2014

From €	To €	Tax Rate %	Tax payable at top of band €
Nil	7,993.46	7.65	611.50
7,993.46	15,980.91	8.50	1,290.43
15,980.91	23,968.36	9.35	2,037.26
23,968.36	31,955.81	10.20	2,851.98
31,955.81	39,943.26	11.05	3,734.59
39,943.26	47,930.22	11.90	4,685.10
47,930.22	55,918.17	12.75	5,703.50
55,918.17	63,905.62	13.60	6,789.80
63,905.62	71,893.07	14.45	7,943.98
71,893.07	79,880.52	15.30	9,166.06
79,880.52	119,757.67	16.15	15,606.22
119,757.67	159,634.83	18.70	23,063.25
159,634.83	239,389.13	21.25	40,011.04
239,389.13	398,777.54	25.50	80,655.08
398,777.54	797,555.10	29.75	199,291.40
Over	797,555.08	34.00	–

Gifts and Inheritance Tax Multiplier

Gifts and Inheritance Tax Multiplier				
Net Worth of Donee		Group I & II Ascendants Descendants Married Couples	Group III 2nd and 3rd degree relatives	Group IV Others
From €	To €			
Nil	402,678.11	1.1000	1.5882	2.0000
402,678.11	2,007,380.43	1.0500	1.6676	2.1000
2,007,380.43	4,020,770.98	1.1000	1.7471	2.2000
Over	4,020,770.98	1.2000	1.9059	2.4000
Some regions have varied the state inheritances tax rates, notably Asturias, Baleares, Cantabria, Cataluña, Galicia, Madrid and Valencia.				

As mentioned, there are no state allowances for lifetime gifts; it is all taxable. However, in practice, day-to-day gifts often go untaxed. It is still very much Spanish practice that when any non-registered property such as furniture, pictures, etc, (ie, most assets excluding property and cars) is given away, no one in Spain would think of filing a gift tax return.

Other state reliefs

MAIN HOME RELIEF

There is a 95% allowance against the inherited value of the main home of the deceased, up to a maximum of €122,606.47 per inheritor, provided the following conditions are fulfilled:

- The inheritor must be a spouse, ascendant or descendant (ie, Group I or II), or, failing that, a more remote relative over the age of 65 who lived with the deceased during the two years prior to their death.

- The property must be retained by the beneficiary for 10 years following the death, but does not need to be the main home of the beneficiary in that period. If the beneficiary dies within 10 years, the succession and gift tax is **not** clawed back. The maximum deduction available is €122,606.47 per inheritor.

Provided the above conditions are fulfilled, the deduction applies regardless of whether the recipient is resident in Spain or not.

BUSINESS RELIEF

There is a 95% reduction in inheritance tax liabilities on family businesses including shares in family companies (those that would be wealth tax exempt) inherited by a spouse, ascendants or descendants. The business or shares must be kept for 10 years following the death of the individual.

DIVORCE

Divorce usually avoids gift tax.

EXAMPLE OF INHERITANCE TAX

Each child inheritor can claim the standard deduction of €15,956,87. For example, if a flat or small house worth €80,000 is left to four children equally, each child inherits €20,000. Each adult child can then deduct the standing deduction of €15,956.87, leaving €4,043.13 each liable to Spanish succession tax. The heirs would pay less than €310 Spanish succession tax each. Note, however, that if the deceased is UK domiciled, the Spanish assets also remain liable to UK inheritance tax, and all assets in the estate of the deceased in excess of £325,000 (2014/15) – or possibly up to twice that if an earlier deceased spouse's nil-rate band is also available – are liable to 40% tax. Any tax paid in Spain can be deducted against the UK tax liability to avoid a double tax charge.

8.4 Cumulation of gifts

Gifts made by the same donor to the same donee within a period of three years, taken from the date each gift is made, are treated as one transaction for gift tax purposes.

The values of all the gifts made in the relevant three-year period are aggregated. The value to take into account is the value of the gift at the time it was made, even if the value has changed since then.

To determine the tax rate applicable, you add up all previous gifts made to the same person within the last three years plus the current gift, and work out what the average rate of tax is on the theoretical total. This average rate is then applied to the latest gift. As a result the rate should increase with each transaction within any three-year period.

8.5 Husband and wife and Spanish inheritance tax

Aside from some recent substantial changes in succession and gift tax for residents of certain regions (see below), there is currently no blanket exemption from Spanish inheritance tax between a husband and wife under the state rules. Where they are both resident in Spain, for example, and the husband dies, then the wife can be fully liable on the worldwide assets she inherits from her husband, subject only to the allowances and reliefs available.

If the couple are living together unmarried, the tax due is even more since the tax would be at least doubled (unless they were registered as *pareja de hecho* and lived in one of the regions that grants the same reliefs to such couples as married couples – currently Cataluña, Andalucía, Islas Baleares and Islas Canarias).

Examples for Couples (round figures):

Assumes heir's estate is under €402,678.11				
Estate €	**Less Exempt €**	**Net €**	**Tax Payable Married €**	**Tax Payable Unmarried €**
50,000	(15,957)	34,043	3,083	9,898
100,000	(15,957)	84,043	9,838	24,831
200,000	(15,957)	184,043	28,250	63,282
400,000	(15,957)	384,043	76,898	162,038
750,000	(15,957)	734,043	180,397	370,288
1,000,000	(15,957)	984,043	262,697	536,245
2,000,000	(15,957)	1,984,043	602,697	1,216,245
5,000,000	(15,957)	4,984,043	1,622,697	3,256,245
10,000,000	(15,957)	9,984,043	3,322,697	6,656,245

* *No €15,957 exemption*

8.6 Interactions between state and autonomous community rules

While Spain sets state succession and gift tax rates and exemptions nationally, the Autonomous Communities (the regions or 'ACs') have the power to vary these in the taxpayer's favour. The ACs are the political and administrative divisions immediately below the state, made up of one or more provinces. For example, Andalucía is an AC of eight provinces; *Comunidad Valenciana* of three; Murcia of one.

8.7 Autonomous community or state rules?

Unfortunately this can be complex. There are not necessarily categorical provisions in the new inheritance reliefs introduced in certain regions that say they will not be available to non-residents, but because of the operation of other underlying rules it is the state (ie, national) and not the regional rules that will apply in the case of inheritance of Spanish property and other assets from non-residents or those who have not been resident in an AC for long enough. State rules will also apply in other circumstances.

Deciding the place where the succession tax has to be paid and the rules that will apply in calculating it is not straightforward and can variously depend upon the AC where:

- In the case of an inheritance, the deceased was habitually resident; *or*
- In the case of a gift of real estate, the property is located; *or*
- In the case of a gift of other assets, the recipient is habitually resident.

This is summarised in the table below.

In the case of inheritances, or gifts of assets other than Spanish real estate, it is the habitual residence of the deceased or donee (as the case may be) which dictates whether the AC rules can apply. People may have moved around Spain, or not have been resident in one AC for long enough since arriving in Spain to be considered habitually resident in a particular AC. In that case, the state rules will apply.

Meaning and importance of habitual residence

In determining habitual residence the last five years up to the date of the taxable event should be considered. An individual will only have one AC of residence for a tax year (or none at all if it cannot be determined). The primary rule is that the AC of habitual residence is where the individual spends most time during the tax year (ignoring temporary absences) and unless proved otherwise will be assumed to be the AC where the habitual home (the *vivienda habitual*) is located. (These rules partly echo the national Spanish tax residence rules.)

The deceased or donee (as the case may be) must have been continuously resident in an AC over the past five years for that particular AC's rules to apply. If the deceased or recipient (as the case may be) was not habitually resident in one particular AC over the past five years, then the AC rules will not apply and the state rules will apply by default. The place of habitual residence has become particularly important recently because of the regional succession and gift tax reforms for those living in Andalucía, Madrid Murcia, Comunidad Valenciana, Islas Baleares and Islas Canarias.

Summary: AC or State Rules

<table>
<tr><th></th><th>Real Estate</th><th>Other Assets</th></tr>
<tr><td>Inheritances</td><td colspan="2">AC where deceased habitually resident, or the state if non-resident or the AC cannot be determined</td></tr>
<tr><td>Gifts</td><td>AC where property located, or the state if foreign real estate</td><td>AC where recipient habitually or the state if non-resident or the AC cannot be determined</td></tr>
</table>

The AC variations only tend to affect inheritances and not lifetime gifts, so if the state rules apply to a gift instead of AC rules, it might not make much difference to the tax. With an inheritance it could make a lot of difference.

8.8 Main variations within the autonomous communities

Important note: Refer to the Summary table above, to determine whether the regional AC rules can apply or not to the transfer in question.

ANDALUCÍA

NOTE

Possible 100% tax relief for Group I and/or II residents on inheritances up to €175,000 (see Other reliefs: possible 100% exemption on page 115).

Personal reductions

Same as state rules.

In Andalucía unmarried couples registered as a *pareja de hecho* (a de facto couple) are within Group II and those in a family unit by way of 'family welcome' *('acogimiento familiar')* in accordance with Andalucía's rules on protecting minor children are included in Groups I or II as applicable, depending on their age.

Main home relief

This is 95% under the same conditions as for the State rules (ie, must be kept for 10 years, maximum reduction of €122,606) and with the additional condition that it must be the main home of both the deceased and the recipient at the time of death.

Other reliefs: possible 100% exemption

No tax is payable provided:

a. The recipient is a close relative falling in Groups I and II (ie, spouse, direct line recipients); *and*

b. The taxable value of the inheritance received is no more than €175,000; *and*

c. The wealth of the recipient is no more than €402,678.

For beneficiaries with a disability of more than 33%, the limit on the taxable inheritance (condition (b) above) is increased to €250,000.

Relief for businesses

Relief of 99% on inheritances of businesses situated in Andalucía provided the business or shares are retained for five years (not 10) following death.

Lifetime gifts to children to acquire first home

There is a 99% relief on the final gift tax payable on monetary gifts to children (<35 years old) of up to €120,000 (€180,000 if descendant has a disability of 33% or more) to purchase their first main home, which must be situated in Andalucía.

The entire cash gift must be used to purchase the main home. The pre-existing wealth of the donee must be less than €402,678. The property must be purchased within one month of the gift and must be noted in the purchase documents.

CATALUÑA

Unmarried couples registered as a *'unión de pareja'* are treated like spouses.

Personal reductions

From 1 January 2014, the personal reductions are as follows:

GROUP I (ie, children under 21)	€100,000, plus extra €12,000 for each year under 21 up to max of €196,000
GROUP II (this has been split into varying degrees of kinship)	Children over 21 €100,000 Spouses €100,000 Other descendants €50,000 Ascendants €30,000
GROUP III (ie, relatives up to the 2nd and 3rd degree, stepchildren and step-parents)	€8,000
GROUP IV	€0 Where the recipient is over 75 €275,000 (this reduction is not cumulative with the disability one)

(This reduction is not compatible with the reduction for disabilities below, ie, you cannot have both).

Reductions for individuals with disabilities:

Disability of between 33% and 65%	€275,000
Disability over 65%	€650,000

It should be noted that both the standard and additional reductions described above are halved where the taxpayer also elects for the following reliefs:

- Reduction for the acquisition of a business or shares in a business.
- Reduction for the acquisition of goods or rights relating to an economic activity.
- Reduction for the acquisition of certain rustic farms for forestry purposes.
- Reduction for the acquisition of assets used in the agricultural business of the deceased.
- Reduction for the acquisition of assets of cultural or natural heritage.
- Any other reduction or exemption where an election is required by the taxpayer or which depends on the taxpayer having to fulfil certain conditions.

The above personal and additional reductions are being brought in gradually in three phases.

Main home relief

Reduction of 95% up to a property value of €500,000. This amount should be pro-rated among the beneficiaries in proportion to each beneficiary's share. Each individual beneficiary's limit cannot be less than €180,000. The property must be kept for five years (not 10) following death. Like the State rule, the relief applies to spouses and children, or relatives over 65 having lived with the deceased for at least two years prior to death.

If the property is sold within the five-year limit, the relief will not be clawed back if the proceeds are used to buy the beneficiary's main home.

If at the time of death the deceased has his effective residence in another home which he did not own, the property which was his main home up to any time during the previous 10 years still qualifies for relief. The 10-year period is not taken into account if the deceased had his last home in a residential care home or health centre.

Other reliefs on death

- 95% reduction for businesses (including agricultural).
- 95% reduction for share in quoted or unquoted companies, provided they are kept for five years following death. Shares in real estate companies are excluded. The participation held must either be 5% individually, or 20% in the family. The deceased, or a member of the family, must have a managing activity in the company.
- 95% reduction on certain assets registered in official registers of cultural heritage, provided they are kept for 10 years following death.
- 100% reduction for beneficiaries of life insurance policies on death up to a maximum of €25,000. The beneficiary must be a spouse, ascendant or descendant, including adopted children.
- 95% reduction for certain rustic farms used for forestry purposes, provided they are kept for 10 years following death.

NOTE

Bank certificates are needed for bank accounts for the year prior to death to present with the inheritance tax return.

New exemption of tax liability

From February 2014, the 99% tax exemption for transmissions between Groups I and II has been removed. Now a decreasing rate is applicable. However, for transmissions between spouses the bonus remains 99% of the tax liability.

Tax Rates and Multipliers
Cataluña – from 1 January 2010

From €	To €	Tax Rate %	Tax payable at top of band €
Nil	50,000	7	3,500
50,000	150,000	11	14,500
150,000	400,000	17	57,000
400,000	800,000	24	153,000
Over 800,000	–	32	–
400,000	800,000	29.75	199,920
Over	800,000	34.00	–

The multipliers are:

GROUPS I and II	GROUPS III	GROUPS IV
1	1.5882	2

Therefore, for spouses, descendants and ascendants, the multiplier does not increase the amount of tax payable.

Gifts tax rates

From 2008 the tax rates for gifts to beneficiaries in Groups I and II (spouses and children) are as follows:

Euro rate total tax payable at top.

ISLAS BALEARES

Varied tax rates and multipliers apply in Islas Baleares (**see Tax rates and multipliers: Islas Baleares** on page 123).

Personal reductions

- For children under 21 who are resident in Spain (not necessarily in the Islas Baleares – for at least one year prior to the date of death), there is a 99% reduction in the inheritance tax payable on death. This is under Art 16 of Law 22/2006 of Islas Baleares.

- For spouses and children over the age of 21, there is a 99% reduction against the taxable inheritance on death, regardless of their residence status. This is under Art 19 of Law 22/2006 of Islas Baleares.

There is a formula for calculating this reduction: Reduction = Tax payable – (taxable inheritance x 0.01).

The effect of the formula is that the tax payable is equal to 1% of the taxable inheritance (rather than 1% of the tax payable as in a) above.

EXAMPLES

Say beneficiary is a child under 21 resident in Spain

Taxable inheritance €100,000.

Tax payable €9,668 (no multiplier applies).

99% reduction (€9,571) (available as deceased resident in Islas Baleares over past five years, Group I beneficiary resident in Spain).

Tax payable €97

Say child over 21 resident in UK

Taxable inheritance €100,000.

Tax payable €9,668 (no multiplier applies).

Reduction = €9,668 - (€100,000 x 0.01 = €1,000) = €8,668. (Even though deceased habitually resident in Islas Baleares over past five years, beneficiary is not resident in Spain).

Tax payable therefore €9,668 - €8,668 = **€1,000**

So, the end result is slightly different, but there is still a huge saving even for beneficiaries resident in the UK.

Unmarried couples within a stable relationship *(las parejas estables)* as defined in *Ley 18/2001* are treated as spouses, provided they are registered in the Community *Registro de Parejas Estables.*

Allowances for inheritances:

GROUP I	€25,000 plus extra €6,250 for each year under 21 up to a maximum of €50,000
GROUP II	€25,000
GROUP III	€8,000
GROUP IV	€1,000

Allowances for individuals with disabilities:

Physical disability of between 33% to 65%	€48,000
Physical disability over 65%	€300,000
Mental disability over 33%	€300,000

Main home relief

100% of the value of the home, up to a maximum of €180,000 with the usual conditions (ie, recipients must be Group I or II and keep the home for at least 10 years after death. Where the recipient is not Group I or II, they must be over the age of 65 and have lived with the deceased for two years before death).

Other reliefs on death

95% of businesses are exempt from wealth tax, provided the beneficiary keeps the business for five years (not 10), and does not carry out any transactions during this time that, directly or indirectly, could substantially reduce the acquisition value of the business.

95% of the value of certain land in protected areas and land in zones of agricultural interest, inherited by recipients in Groups I and II.

100% reduction for beneficiaries of life insurance policies on death up to a maximum of €12,000. The beneficiary must be a spouse, ascendant or descendant.

If the same asset is subject to two or more inheritances within a 12-month period, the amount of tax paid on the previous inheritances can be deducted from the taxable base *(base imponible).*

Reliefs for gifts

For spouses and children, there is a 93% reduction in the tax payable on certain lifetime gifts.

There is a 57% deduction for cash gifts to descendants under the age of 36 or with a physical disability of more than 65% or a mental disability of more than 33% to buy their first main home, up to a maximum of €60,000 or €90,000 if disabled. The income of the donee must not exceed €18,000 and their wealth must not exceed €400,000. The value of the property must not exceed €180,000 and the surface area must not exceed 120m^2. The donee must reside in the property for a minimum of three years from the date of acquisition. The gift must be registered as such and the property must be bought within six months of the gift (or the first gift if a series of gifts is made).

There is also a 57% deduction for cash gifts to descendants under the age of 36 to purchase their own company or business or shares in a company, up to a maximum of €30,000 or €42,000 for individuals with a disability of more than 33%. The gift must be registered in the *escritura pública* and the business or shares must be purchased within six months of the gift. The donee's wealth must not exceed €400,000. The annual turnover must not exceed €3m in the case of a company, or €1m in the case of a business. If shares are being purchased, they must represent at least 50% of the total shares of the company and the donee must carry out management functions within the company.

There are also 99% exemptions on gifts for the purpose of buying a business, or where a business is inherited provided certain conditions are met.

Tax rates and multipliers: Islas Baleares

From €	To €	Tax Rate %	Tax payable at top of band €
Nil	8,000	7.65	612
8,000	16,000	8.50	1,292
16,000	24,000	9.35	2,040
24,000	32,000	10.20	2,856
32,000	40,000	11.05	3,740
40,000	48,000	11.90	4,692
48,000	56,000	12.75	5,712
56,000	64,000	13.60	6,800
64,000	72,000	14.45	7,956
72,000	80,000	15.30	9,180
80,000	120,000	16.15	15,640
120,000	160,000	18.70	23,120
160,000	240,000	21.25	40,120
240,000	400,000	25.50	80,920
400,000	800,000	29.75	199,920
Over	800,000	34.00	–

Gifts and Inheritance Tax Multiplier				
Net Worth of Donee		Group I & II Ascendants Descendants Married Couples	Group III 2nd and 3rd degree relatives	Group IV Others
From €	To €			
Nil	400,000	1.1000	1.5882	2.0000
400,000	2,000,000	1.0500	1.6676	2.1000
2,000,000	4,000,000	1.1000	1.7471	2.2000
Over	4,000,000	1.2000	1.9059	2.4000

ISLAS CANARIAS

Personal reductions

The rates below have are with effect from 1 July 2012.

Allowances for inheritances:

GROUP I	Children aged under 10 years – 100% deduction up to a maximum of €138,650 Children aged 10 to 15 years – 100% deduction up to a maximum of €92,150 Children aged 15 to 18 years – 100% deduction up to a maximum of €57,650 Children aged 18 to 21 years – 100% deduction up to a maximum of €40,400
GROUP II	Spouse – €40,400 Children (over 21 years) – €23,125 Other ascendants and descendants – €18,500
GROUP III	€9,300
GROUP IV	€0

Additional deductions for disabled:

33% - 65% disability	€72,000
65% or more	€400,000
Individuals aged 75 years or more	€125,000

(This reduction is not compatible with the reduction for disabilities, ie, you cannot have both).

Unmarried couples registered as a *pareja de hecho* are treated as spouses.

Children who are in the family unit by way of 'family welcome' *('acogimiento familiar')* are treated as adopted children in Groups I and II as applicable, depending on their age.

Main home relief

99% deduction up to a combined maximum of €200,000 with the usual conditions, but the property needs only be kept for five years (rather than 10 years as per state rules).

Relief for family business

99% deduction provided spouse or children (including adopted) inherit, or relatives up to 3rd degree where there are no children, they keep the business for five years, the value of the company is no more than €3m (or €1m if an individual business), and the shares were exempt from wealth tax in either of the two years preceding death.

Other reliefs on death

99% for life assurance policies up to a maximum €23,150 provided the recipient is a spouse (or partner), ascendant, descendant or adopted.

Reliefs for gifts to buy a main home or to a disabled descendant

85% deduction for cash gifts to descendants under 35 with their habitual residence in the Canaries to buy their main home up to a maximum of €24,040 or 90% reduction up to €25,242 if disability of 33%, or 95% reduction up to €26,444 if a 65% disability. The gift must be registered as such in *escritura pública* and the property must be bought within six months of the gift (or the first gift if a series of gifts is made). The property must be situated in the Canaries and the recipient must live in the property as his main home for at least five years from the date of acquisition. If the property is to be built or renovated, work must be started within six months of the gift and must be finished within two years of the date the work started.

MURCIA

Personal reductions

Same as state.

But, for beneficiaries under the age of 21 who are in Group I (ie, descendants and adopted children under the age of 21) there is a 99% deduction of the taxable inheritance. This is applied after all other deductions.

For 2012 and 2013, for Group II beneficiaries (descendants and adopted over 21, spouses and ascendants), the taxable inheritance is reduced by 99%, provided the taxable value of the inheritance received (after state deductions and allowances) is less than €300,000 per beneficiary (or €600,000 if disability of 65% or more).

Main home relief

Same as state rules.

Relief for family business

99% for business or personal companies. The income from the business must represent at least 50% of the deceased's total professional and employment income, and the deceased must have owned at least 10% of the business (or 20% together with spouse or relatives up to 3rd degree – ie, Group III), and exercised managerial functions within the business. The business needs to be kept in Murcia for at least five years after death. The relief specifically does not apply to property investment businesses.

Relief for gifts of business on retirement

99% of business or shares in a company situated in Murcia gifted to spouses or children, provided donor is over 65 or permanently incapacitated. Specifically does not apply to property investment companies. The donor must carry out management functions within the business and the income from the business must represent at least 50% of the donor's total professional and employment income. The business must be kept by the donee in Murcia for at least 10 years.

Gifts of future main home

99% for gifts of property situated in Murcia which will become the recipient's main home or cash gifts to buy the recipient's main home, up to €150,000. Above this amount a fixed rate of 7% will apply. The donee must be within Groups I or II, the gift must be registered and the donee must not have another main home available at the time the gift is made.

Gift of business

99% for gifts to set up or acquire a personal company or business in Murcia up to €100,000 (€200,000 if the donee has a disability of 33% or more). The donee must be under 35 and within Groups I or II, the gift must be formalised and the business or company must be set up or purchased within six months of the date of the gift. The donee's wealth must be less than €300,000 at the date of the gift. In the case of a purchase of a business or company, turnover for the previous accounting period must not exceed €3m for a company or €1m for a business. If shares are being purchased, they must represent at least 50% of the share capital of the company and the donee must exercise a managerial role within the company. The investment must be retained for at least five years.

COMUNIDAD VALENCIANA
(Valencia, Alicante and Castellon Provinces)

Varied tax rates and multipliers apply in the *Comunidad Valenciana* – (see **Tax rates and multipliers: *Comunidad Valenciana*** on page 131).

NOTE

Possible 75% tax relief for Group I and/or II residents – **(see 75% Tax Reliefs on page 130).**

Personal reductions

On death:

GROUP I	€100,000, plus extra €8,000 for each year under the age of 21 up to a maximum of €96,000
GROUP II	€100,000
GROUP III	€100,000
GROUP IV	Nil

Reductions for individuals with disabilities:

Physical disability of between 33% to 65%	€120,000
Physical disability over 65%	€240,000
Mental disability over 33%	€240,000

On gift:

GROUP I (children)*	€40,000, plus extra €8,000 for each year under the age of 21 up to a maximum of €96,000
GROUP II (children and parents)*	€40,000

**The gift reliefs between parents and children are extended to the relevant grandparents and grandchildren where a parent has died.*

In both cases, the donee must be habitually resident in the *Comunidad Valenciana* at the date of the gift and the wealth of the donee must not exceed €2m. The gift must be officially registered. All gifts within the previous five years are taken into account.

The above deductions are not given where the donor himself has benefited from these deductions in the previous 10 years, or where the donee has gifted assets (to someone other than the current donor) that have benefited from this deduction within the last 10 years.

Physical disability of 33%	€120,000, plus above deductions to 65%
Physical disability over 65%	€240,000, plus above deductions
Mental disability over 33%	€240,000, plus above deductions

Gifts within the past five years are taken into account for the above deductions.

Main home relief

Same as state rules.

Relief for family business

95% with the usual conditions, but shares only need to be kept by recipient for five years. The business activity must be carried out by the donor in a habitual, personal and direct manner, and the income from the business must represent the donor's principal source of income.

Other reliefs on death

Variable reduction – 25%, 50% or 95% on assets which are registered in the official national or the official Valencia register of cultural heritage.

Reliefs for gifts of businesses

95% for businesses gifted to spouses or children (or to parents if there are no children) provided it is kept for five years where the donor is 65 or over. The reduction is 90% where the donor is 60-64 at the date of the gift.

75% Tax Reliefs

There is 75% reduction in the Succession and gift tax payable on:

- Inheritances by Group I and II relatives who are habitually resident in the Valencian Community at the time of the deceased's death.
- Inheritance by any relative who is physically or mentally disabled (the level of disability is defined as 65% or more physical; 33% or more mental), regardless of the degree of relationship with the deceased. There is no mention of a residence requirement so the standard rules for deciding whether the State (national) or regional rules apply will apply (**see Section 7.8**).
- Gifts to children and parents (including adopted/ive) – but not more remote ascendants/descendants[2] and not spouses – whose existing wealth is below €2m, subject to a limit of €150,000 per beneficiary. Again, the beneficiaries must be habitually resident in the Valencian Community at the date of the gift. All gifts from the same donor during the five years immediately beforehand are taken into account.

The relief does not apply:

- Where the current donor himself has benefited from the 75% (previously 99%) reduction on the receipt by him of the same assets or others of equivalent value in the past 10 years; *or*
- Where the current beneficiary has made a gift to a different donee (ie, other than the current donor) of other assets of equivalent value in the past 10 years that has also benefited from the reduction.
- Except on any new amount in excess of the earlier donation (up to the relievable limit).

[2] From 1 January 2008, the gift relief between parents and children is extended to the relevant grandparents and grandchildren where a parent has died.

In other words, there cannot be sequential 99% relief on the same assets or equivalent value within 10 years of either the current donor receiving a 99% relieved gift, or the current donee having made a 99% relieved gift to someone else.

Lifetime gifts to children and parents (including adopted/ive) who are physically or mentally disabled (as defined above). The standard rules will apply in deciding whether State or regional reliefs apply.

Tax rates and multipliers: ***Comunidad Valenciana***

From €	To €	Tax Rate %	Tax payable at top of band €
Nil	7,993.46	7.65	611.50
7,993.46	15,662.38	8.50	1,263.36
15,662.38	23,493.56	9.35	1,995.58
23,493.56	31,324.75	10.20	2,794.36
31,324.75	39,155.94	11.05	3,659.70
39,155.94	46,987.13	11.90	4,591.61
46,987.13	54,818.31	12.75	5,590.09
54,818.31	62,649.50	13.60	6,655.13
62,649.50	70,480.69	14.45	7,786.74
70,480.69	78,311.88	15.30	8,984.91
78,311.88	117,407.71	16.15	15,298.89
117,407.71	156,503.55	18.70	22,609.81
156,503.55	234,695.23	21.25	39,225.54
234,695.23	390,958.37	25.50	79,072.64
390,958.37	781,916.75	29.75	195,382.76
Over	781.916.75	34.00	–

Gifts and Inheritance Tax Multiplier				
Net Worth of Donee		Group I & II Ascendants Descendants Married Couples	Group III 2nd and 3rd degree relatives	Group IV Others
From €	To €			
Nil	390,657.87	1.1000	1.5882	2.0000
390,657.87	1,965,309.58	1.0500	1.6676	2.1000
1,965,309.58	3,936,629.28	1.1000	1.7471	2.2000
Over	3,936,629.28	1.2000	1.9059	2.4000

MADRID

Varied tax rates and multipliers apply in Madrid – (see **Tax Rates and Multipliers-Madrid** on page 134).

NOTE

There is a 99% reduction in the tax payable for inheritances and gifts received by Group I and II beneficiaries (spouses, ascendants and descendants including adopted). Gifts must be registered and if made in cash or by bank transfer their source must be justified and must be noted in the public document.

Personal reductions

On death:

GROUP I	€16,000, plus extra €4,000 for each year under 21 up to max of €48,000
GROUP II	€16,000
GROUP III	€8,000
GROUP IV	Nil

Reductions for individuals with disabilities:

Physical disability of between 33% to 65%	€55,000
Physical disability over 65%	€153,000

Individuals within a *unión de hecho* are treated as married couples provided they meet the conditions set out in *Ley 11/2001* of 19 December on *Uniones de Hecho de la Comunidad de Madrid.*

Main home relief

95% up to €123,000 on inheritances by a spouse, ascendants or descendants, or a relative over 65 who has lived with the deceased for two years prior to death. The property must be kept for five years.

Other reliefs on death

100% of the amounts received from life insurance contracts, up to €9,200, where the recipient is a spouse, ascendant of descendant, including adopted.

95% of a business, personal company or shares in a company exempt from wealth tax where the beneficiary is a spouse or descendant (including adopted), of the deceased. Where there are no descendants the reduction can apply to relatives up to the 3rd degree. The business or company must be kept for five years.

Tax Rates and Multipliers – Madrid

From €	To €	Tax Rate %	Tax payable at top of band €
Nil	8,313.20	7.65	635.96
8,313.20	16,001.35	8.50	1,289.45
16,001.35	24,002.01	9.35	2,037.51
24,002.01	32,002.70	10.20	2,853.58
32,002.70	40,003.36	11.05	3,737.66
40,003.36	48,004.04	11.90	4,689.74
48,004.04	56,004.71	12.75	5,709.82
56,004.71	64,005.39	13.60	6,797.92
64,005.39	72,006.05	14.45	7,954.01
72,006.05	80,006.73	15.30	9,178.12
80,006.73	119,947.58	16.15	15,628.56
119,947.58	159,888.45	18.70	23,097.51
159,888.45	239,770.16	21.25	40,072.37
239,770.16	399,408.59	25.50	80,780.17
399,408.59	798,817.20	29.75	199,604.23
Over	798,817.20	34.00	–

Gifts and Inheritance Tax Multiplier				
Net Worth of Donee		Group I & II Ascendants Descendants Married Couples	Group III 2nd and 3rd degree relatives	Group IV Others
From €	To €			
Nil	403,000	1.1000	1.5882	2.0000
403,000	2,008,000	1.0500	1.6676	2.1000
2,008,000	4,021,000	1.1000	1.7471	2.2000
Over	4,021,000	1.2000	1.9059	2.4000

8.9 Spain's succession tax exemptions against EU law

In certain Autonomous Communities (AC), there can be almost total exemption from Spanish succession and gift tax in relation to transfers to descendants and spouses, where the deceased was habitually resident in that region over the last five years.

However, the AC exemptions are not applicable where the deceased was not resident in Spain except where expressly provided for.

The higher tax burden imposed on non-Spanish residents is considered discriminatory by the European Commission (EC).

The EC first sent Spain a 'reasoned opinion' about these discriminatory rules on 5 May 2010. While Spain tweaked its laws, they are still not fully compliant with EU law. The Commission therefore sent a 'complementary reasoned opinion' to Spain on 16 February 2011, stating that Spain's inheritance and gift tax rules are in breach of the EU Treaty's rules on the free movement of workers and capital. Spain had two months to respond.

On 27 October 2011, the European Commission decided to refer Spain to the ECJ for its discriminatory rules on inheritance and gift tax that require non-residents to pay higher taxes than residents. The Commission had already formally requested Spain on 5 May 2010

(IP/10/513) and additionally on 17 February 2011 to take action to ensure compliance with the EU rules with regard to inheritance and gift tax provisions.

However, no amendments have been made to the Spanish legislation on the matter.

The Commission considers this discriminatory tax treatment as an obstacle to the free movement of people and capital, and is in breach of Articles 45 and 63 of the EU Treaty.

It can reasonably be expected that the ECJ will decide that the rules at issue are indeed in breach of EU law. However, such a decision by the ECJ would not directly improve the tax situation of EU tax residents who have paid inheritance taxes at a higher rate than Spanish tax residents.

Should EU tax residents want to challenge the excess inheritance tax paid in Spain, they will have to personally claim for a refund of the excess from the Spanish tax authorities now. If not, these protective claims could be time barred by the time the ECJ judgement is given.

If the outcome of a personal refund claim process is successful, the claimant will be entitled to a refund of the tax plus late payment interest.

At the time of writing, nothing has yet been heard about this.

8.10 Responsibility to declare the death to the *Hacienda*

On the first death, the surviving spouse has six months to inform the authorities and pay the tax, with an extension to pay the tax of up to six months by application within the first five months. The extension is deemed given if no reply is received within one month. Note that if this is not done, and if the title deeds *(escritura)* to a property remain in the deceased's sole, or joint, names then it cannot be sold. There are also heavy penalties and interest for non-payment of the Spanish succession and gift tax on time.

8.11 Charities

Succession and gift tax applies only to individuals and therefore charities cannot be subject to Spanish succession and gift tax.

Usually, if a gift or inheritance is made to a Spanish company or other legal entity, that entity pays Spanish corporation tax on the amount received. However, charities benefit from a special tax regime in Spain, along with other non-profit making associations such as NGOs, sporting federations, the Spanish Olympic Committee, Red Cross, etc. Specific conditions have to be met for the charity to qualify. These conditions centre mainly around the pursuit of activities of general interest, and the requirement to use 70% of their income for such activities. Income received by a qualifying charity by way of gift or inheritance is exempt from corporation tax. Only income arising from certain economic activities is taxable on the charity.

8.12 Overseas taxes on gifts or inheritances

Spain only has double tax treaties for inheritance and gift tax purposes with three countries: France, Greece and Sweden. In the absence of a treaty, normal Spanish rules will apply to any gift or inheritance but where any overseas gifts or inheritance tax is actually paid, this overseas tax is deductible against any Spanish tax liability on the same transfer. This is a Spanish unilateral relief for overseas gift and succession and gift tax. However, if the overseas tax rate is higher than the Spanish rate the Spanish authorities will, of course, not provide a refund of the difference.

8.13 Common law spouse

Spanish state law does not recognise a common law spouse. They have no standard deduction of €15,956.87, and is taxed at the highest rate on inheritances (double the rates which apply for a spouse) as they are treated as a non-relative – but see below re *pareja de hecho* and same-sex marriage. Sometimes, the best and simplest advice we can give in these circumstances is to consider marriage.

PAREJA DE HECHO

In most regions in Spain, unmarried partners (including those of the same sex) can register in the local Register Office as a *pareja de hecho* (also called *pareja estable, unión de pareja, unión de hecho* in certain regions). This gives unmarried partners the same rights under local law as a married couple living together. For **succession and gift tax purposes,** the registration as a *pareja de hecho* is currently only effective in Andalucía, Cataluña, Canaries, Baleares, Madrid, Asturias, Castilla y León and (allowances only) Cantabria. In these ACs couples registered as a *pareja de hecho* (or equivalent) will be treated as married couples for succession and gift tax purposes, and will be entitled to the same rates and allowances as married couples (except in Cantabria where they only receive the same allowances as married couples, not the same rates). However, in all other regions, couples registered as a *pareja de hecho* are still treated as 'strangers' for succession and gift tax purposes.

8.14 Same sex relationships

Same-sex marriages became legal in Spain on 4 July 2005. Couples who marry are entitled to the same tax treatment (for all taxes) as heterosexual married couples, ie, they may file joint income tax returns and are entitled to the same succession and gift tax exemptions. They are also subject to Spanish succession laws (but see below for UK nationals) and will be entitled to Spanish legal residence status (if one of the parties is a Spanish citizen).

It is worth noting that with the rights come responsibilities, and therefore, if a gay marriage breaks down, child support payments will be mandatory in the case of divorce.

It appears that marriages between Spanish resident UK nationals are permitted (and will be recognised in the UK as a civil partnership).

For two UK nationals not resident in Spain, the position is not quite so clear. Spain generally allows an individual's national law to override Spanish law in certain areas (such as succession law – see below), and this may continue. However, it is uncertain whether same-sex couples will be able to visit Spain to marry, as a requirement of marriage in Spain is that one or both of the parties be resident in Spain.

8.15 The usufruct

Introduction

One method of reducing Spanish succession tax is to leave the surviving spouse a 'usufruct', which is a life interest, in the family house, rather than leaving them your half of the property absolutely.

By leaving the spouse a usufruct, the underlying ownership *(nuda propiedad)* of the house can pass to the children. The *usufructario* (the life interest holder) can live in the property free for the rest of their lifetime. This device can save a lot of Spanish inheritance tax.

Life interests (usufructs) and reversionary interests

The ownership of any asset in Spain can actually be divided into two parts: a 'life interest' *(usufructo)* and a 'reversionary interest' *(nuda propiedad)*. For example, consider the ownership of a villa. Mr Smith could own the entire villa outright, and then decide to give his only son the reversionary interest in the villa, while creating a 'life interest for his (ie, Mr Smith senior) own life'.

Thus, Mr Smith senior does **not** now own the villa outright; instead, he owns the right to live in the villa rent-free for the rest of his life. Similarly, his son, while owning the reversionary right to the villa, does so subject to the fact that his father, who owns the life interest in the villa, has the legal right to remain in the villa for his lifetime. When Mr Smith dies, the son's reversionary right will no longer be subject to the life interest, and so the son in effect will then finally own the villa outright. In this example, Mr Smith had only one child. Had he had three children, he could have passed over the reversionary interest to the three of them equally if he so wished.

Because the transfer of a reversionary interest is a gift, it has to be valued for tax purposes. There may be some Spanish gift tax payable on the creation of the reversionary interest (note that there are no exempt limits on a gift in one's lifetime). How does the tax authority value such an interest? You can see that a reversionary interest of a villa with a 103-year-old life tenant is much more valuable than the same villa with a 48-year-old life tenant.

Age of life tenant:		60	
Deduct		(20)	This is always 20
	'A'	40	
Base		70%	
Less	'A'	(40)	This is the figure calculated above
		30%	

30% of the freehold value (€80,000) is €24,000.

The formula has now calculated the value of the life interest as €24,000. By simple arithmetic, the value of the reversionary interest must be €56,000 (€80,000 less €24,000).

There are two exceptions to the application of the formula. Firstly, however old the life tenant, the value of the life interest can never be less than 10% of the open market vacant possession value. Consequently, the value of the reversionary interest cannot exceed 90%. Secondly, if the life tenant is less than 20 years old, the value of the life interest cannot exceed 70%, and consequently the value of the reversionary interest cannot be less than 30%.

What happens if a surviving spouse with a usufruct wants to sell the property? The surviving spouse would only have a life interest in the property. Theoretically, this could be sold as an independent transaction. The new purchaser would buy the right to remain within that property as long as the surviving spouse lives. As you can imagine, there is not much of a market for this asset.

The children could independently sell what we would call the freehold. In Spain, they refer to this as the bare ownership *(nuda propiedad)*. There is a better market for such an asset, as it is an appreciating one, whereas the surviving spouse's life interest is a depreciating asset.

In practice, the children and the surviving spouse would sit down together and agree between themselves what to do. They would probably sell the entire property, and purchase a new one with the children continuing with the freehold interest and the surviving spouse continuing with the life interest. As you can see, however, the surviving spouse cannot make such a decision independently of the children. They will require the children's acceptance. If the children are under 18 years of age, the surviving spouse would need to seek permission from a judge to sell the property.

Tax savings: examples

Based on the concept of the usufruct, here are three different ways of how a Spanish house might be left to one's heirs, and how tax might be saved. The examples assume that the house is the only asset of the estate. The possible 95% main home reduction is not taken into account in this example.

EXAMPLE 1

House in husband's name – tax payable €6,809.56

Assume that the house is worth €80,000 and is in the husband's name. Assuming the husband dies, leaving the home to his wife, the Spanish inheritance tax payable (ignoring all other assets) is:

Value of house	€80,000
Group II deduction	(€15,956.87)
Taxable	€64,043.13
Tax payable	**€6,809.56**

EXAMPLE 2

House in joint names – tax payable €2,043.05

If, instead the house was in joint names, then the calculation is different:

Value of 50% interest	€40,000
Group II deduction	(€15,956.87)
Taxable	€24,043.13
Tax payable	**€2,043.05**

EXAMPLE 3

House in joint names with usufruct and three children – no tax

If, instead, the house was in joint names but the husband left his wife with a life interest, after which time it passed to the three children equally, then there is no tax to pay. In this example, on his death the husband leaves:

- A life interest in 50% of the house to his wife.
- The title to that same 50% to his children (subject to the wife's life interest).

If the wife is 59 years old, the value of the wife's life interest is 30%, and the children's reversionary interest is 70%. The tax calculation is:

	Wife	Children
Interest left	Life Interest	Reversionary
Respective values	30%	70%
50% value: €40,000	€12,000	€28,000
Spouse deduction	(€15,956.87)	–
Children's deductions	3 x €15,956.87	(€47,870.61)
Taxable	Nil	Nil

Conclusion: home ownership

Thus, by the husband and wife having the home in joint names, and by their Spanish Will leaving a life interest to the other, with the ultimate ownership to the children equally, there is a Spanish inheritance tax saving.

If a couple purchases a property jointly, and one wants to transfer their half of the property to the other, it is not necessary to actually do so in the conventional method. Instead, the couple can dissolve the 'community' of ownership, keeping the property in the name of just one of them, and thus paying a document fee of only 0.5% (1% in Andalucía), instead of property transfer taxes of 6 to 8%.

"What is the difference between a taxidermist and a tax collector? The taxidermist takes only your skin."

MARK TWAIN

Spanish Inheritance and Gift Tax

Spanish inheritance and gift tax – and the implications of UK inheritance tax – are important to consider but the key is how they may impact on your personal situation and what you and do limit their impact.

We need to talk.
www.blevinsfranks.com

CHAPTER 9

Local Spanish Taxes

Both residents and non-residents pay these taxes. The taxes payable fall into various categories.

9.1 *Impuesto sobre bienes inmuebles (IBI)*

The *IBI* is paid if you own a residential property and use it yourself (or have it available for your use). It is paid by the person who occupied the property on 1 January in any year. It is not usually apportioned if they later move.

The tax is raised and spent by the Town Hall *(ayuntamiento)* of the area where you live.

It is calculated on the basis of the notional rental value *(valor catastral)* of your property. You can appeal against the valuation decision, but the sums involved are usually so small it is not worthwhile.

The amount you will be charged will be the rental value multiplied by the tax rate fixed in your locality.

If the *IBI* is not paid by the previous owner of a second-hand property, the local Town Hall has the power to charge the new owner for unpaid *IBI* for up to a maximum of five years arrears (and in addition penalties of up to 20%). When buying a property, you should ask to see the latest *IBI* receipt to confirm that this has been paid, and it may be a sensible precaution to insist upon seeing the receipts for the previous five years, in order to ensure that you are not liable to pay any arrears. When selling a property, you will need the latest *IBI* receipt, and as a sign of good faith, you may wish to show receipts for the previous five years. You should also confirm that the ownership information has been changed at the registry, or you may continue to be liable for the *IBI* on a property you no longer own.

Also, the current *IBI* receipt will be required by the notary, as it contains the *catastral* reference number.

9.2 *Plusvalía*

This is a local tax levied by Spanish Town Halls on the increase in the value of urban land (ie, excluding the value of any buildings) when it is sold or otherwise changes hands. Its full name is *Impuesto Sobre el Incremento de Valor de los Terrenos de Naturaleza Urbana* but is abbreviated to *Plusvalía* which literally means 'capital gain'. The real growth in value of the land is reflected by the allowance of a realistic allowance for inflation, unlike the token allowance in Spanish mainstream capital gains tax.

Some transactions are exempt, including transfers between husband and wife, changes in ownership within the immediate family due to court judgements in matrimonial proceedings, and certain technical ownership changes involving rights of way, etc. However, it remains payable on the disposal of your main residence, even if you are Spanish tax-resident.

The tax rate varies depending upon the size of the local population and the length of ownership. For a town of more than 100,000 inhabitants the minimum tax rate is 20% and the maximum 30%, with the Town Hall fixing a rate within this.

Because of the increase in land values over the past 20 years, despite the realistic allowance for inflation, the amount of tax payable tends to be much higher the longer the property has been owned. The inflation-enhanced cost is still only likely to be a small figure compared to the current value if it has been owned for 20 years rather than five years, say.

A declaration must be made by the vendor (or purchaser) within 30 days, but in practice the Town Hall usually calculates the tax based on the property values entered in the local property register from the deeds of sale, and this figure is obtained by the notary (or *gestor*) appointed by the vendor (or purchaser) in order that they can make the required declaration.

9.3 Taxes on property transfers

Spanish VAT *(IVA – Impuesto sobre el Valor Añadido)* at 10% is payable on new build properties (eg, acquired from a developer), or resold, unregistered properties (ie, an off-plan property acquired from another individual prior to being registered).

There is no VAT in the Canary Islands; instead, a sales tax (the *IGIC*) of 5% is levied.

In addition to *IVA*, a document fee – *Impuesto sobre Actos Jurídicos Documentados (AJD)* – is levied, which is similar to UK stamp duty, of 0.5% (1% in Andalucía).

Impuesto sobre Transmisiones Patrimoniales (ITP) is Spanish property transfer tax (similar to the UK stamp duty land tax) levied on all other residential property acquisitions (ie, once the property has been registered). *ITP* is generally levied at 7% of the purchase price, although some regions have increased it to as much as 10%.

9.4 Refuse tax *(Basura)*

Rubbish collection charges are, in some areas, raised separately.

9.5 Other local taxes

Town Halls can also raise taxes for other projects and to cover shortfalls.

9.6 Payment of local property taxes

A demand for payment is sent each year. The sum claimed must be paid by the specified date (which varies from place to place). Failure to do so incurs a 20% penalty. It is probably simplest to arrange for payment from your bank by direct debit.

9.7 VAT

The general VAT rate is 21%, and the reduced rate 10%.

CHAPTER 10

Tax Administration

10.1 Spanish tax office

The Spanish tax office is called the *Hacienda,* and is part of the *Ministerio de Economia* y *Hacienda.* Each province has a head office (called *delegación*) with satellite offices (called *administración*). For example, the Cádiz and Málaga provinces have the following offices:

Delegación	Cádiz	Málaga
Administraciones	San Roque Marbella Chiclana Puerto Santa María Sanlucar de Barrameda Ubrique	Málaga Algeciras Málaga Oeste Vélez Málaga Antequera Alora Ronda

For a non-Spaniard, the tax reference number is known as *NIE (Número de Identificación de Extranjero).* Tax returns must be presented between 1 May and 20 June. Residents can choose to make a joint income tax return as a family unit or to file individual returns.

10.2 Tax havens

The following countries are considered tax havens under Spanish law:

Anguilla
Antigua and Barbuda
Bahrain
Brunei
Dominique
Fiji
Gibraltar
Grenada
Guernsey
Jersey
Jordan
Lebanon
Liberia
Liechtenstein
Macao
Malvinas Islands
Isle of Man
Marian Islands
Mauritius
Monaco
Montserrat
Nauru Republic
Oman
Soloman Islands
Seychelles
Turks and Caicos Islands
Vanuatu
British Virgin Islands
US Virgin Islands

10.3 Non-declaration of tax

It is the responsibility of the individual to complete their tax return correctly. A Spanish tax resident is liable to tax on worldwide income and gains. Unlike the UK, where the tax authority can go back 20 years, the Spanish tax authority can only go back four years from the date the tax return should have been submitted. If, for example, a Spanish tax return must be submitted by 30 June for the previous calendar year, then on 29 June 2017, the tax authorities can go back to the return that should have been submitted by 30 June 2013, ie, income year ended 31 December 2012. Two days later, on 1 July 2017, they can only go back to 30 June 2014 (income for the year ended 31 December 2013).

This can be summarised as:

Date	How Far Back?
01/07/2014	Tax year ending 31/12/2010
30/06/2014	Tax year ending 31/12/2009
01/07/2013	Tax year ending 31/12/2009
30/06/2013	Tax year ending 31/12/2008

In Spain, there are two types of tax offence: a normal offence, and a more serious one. Serious offences are graded depending on whether or not the offence has been repeated, resistance or obstruction has occurred, fraudulent methods have been used, information has been hidden, whether the amounts involved are relatively large, or there has been a lack of co-operation or delay. As well as interest due on overdue tax, the percentage penalties range from 50% to 150% of the amount unpaid and/or undeclared. As in the UK, the tax penalty can be reduced by co-operation with the investigation.

10.4 Enforcement overseas and right to information

The Spanish tax authorities can block bank accounts in Spain and block the possible sale of a property in Spain. They cannot normally enforce the collection of tax against assets outside Spain although, within the EU there is now more co-operation between the legal enforcement systems than ever, and many tax treaties now include exchange of information and mutual assistance clauses. The Spanish tax authorities may receive information from third parties in respect of an individual's tax position. The taxpayer has the right to obtain this information from the tax authorities. This enables the taxpayer to check the information before completing his income tax declaration. If the taxpayer considers that some of the information given by the tax authorities is wrong, they can, of course, have it altered. Many individuals are now taking advantage of this right to obtain information in order to see exactly the data held by the tax authorities before completing their tax return.

10.5 Penalties on tax retentions

As there are a number of penalties applicable to incorrect withholding tax, Spanish financial institutions will usually raise queries, and are likely to refuse to make payments without withholding tax unless they have sufficient information to establish that a retention of tax is not required. Thus, a bank is likely to refuse to carry out instructions from a Spanish taxpayer to transfer funds to an offshore tax haven unless a tax retention is made, or the taxpayer can provide adequate assurance that the withholding tax has either already been paid or is not required. Similarly, the taxpayer must provide information to a Spanish financial institution about the receipt of money from abroad in excess of €30,005.06 from non-Spanish entities.

Receipts include transfers of funds received and cheques banked from a non-Spanish bank account. The financial institution is required to pass on such information to the Bank of Spain, and it is then likely to be forwarded to the tax authorities.

10.6 Offshore companies/entities in tax havens

Where an individual has control, either directly or indirectly, of an overseas company then the income and gains within that company will be imputed on the Spanish resident taxpayer, and the individual taxed accordingly. Where a company is based in a tax haven, it is assumed in Spanish tax law that the income arising from such entities is 15% of the original cost, unless proof is provided to the contrary.

3% Penalty tax on offshore companies owning property in Spain

A 3% tax charge *(Gravamen Especial sobre Bienes Inmuebles de Entidades no Residentes)* applies to non-Spanish resident entities that own immovable property in Spain or rights of use or enjoyment over such property.

The tax is calculated as 3% of the *catastral* value of the property.

Where a non-resident entity owns a share of the property together with other individuals or entities, the 3% tax is payable only on the share of the property that is owned by the non-resident entity.

10.7 Exchange of information

As from 6 April 2001, details of ALL interest bearing bank or building society accounts in the UK, with a Spanish address, will be sent by the UK Inland Revenue to the Spanish tax authorities.

In fact, this information exchange is with over 30 other countries, and not just Spain.

In June 2002, the Spanish authorities announced that they had signed a similar agreement with the Channel Islands. This follows the discovery of secret bank accounts held by *BBV (Banco Bilbao Vizcaya)*, and effectively ends the Channel Islands banking secrecy rules.

These agreements are in addition to the European Savings Tax Directive, which came into force on 1 July 2005 and which affects all countries which have signed up to it, including the EU Member States (including all states when they join the EU) and also Switzerland, the Isle of Man, the Channel Islands, the Cayman and British Virgin Islands.

From March 2011, bank and credit institutions in Spain must automatically report to the *Hacienda* all cash payments or receipts exceeding €3,000 (whatever the currency).

Previously, credit institutions had to report transactions by cheque where the amount exceeded €3,000 but only cash transactions above €100,000.

While the Spanish authorities could previously obtain such information from the banks by requesting it specifically, now the banks are obliged to provide the information automatically each year. The banks will also report the individual's (or company) name, *NIF* number, date of transaction, if it is a receipt or payment, the amount in Euros, and the relevant bank account number.

The aim is for the Spanish tax authorities to be better able to detect tax evasion.

10.8 Example of exchange of information

In 2001/2002, the German tax authorities investigated German citizens with properties in Spain. The procedure was as follows:

The German authorities asked their Spanish counterpart for a list of properties owned by Germans. They got it, in accordance with international agreements (OECD and EU). The Spanish side would have had no problems in making such a computer list, since the names and nationalities are contained in the title deeds registered in the central property register (in Madrid); they are on the list of the *catastro* (office for valuation of property for fiscal purposes); they are on the lists of the *Banco de Espana* (Spanish national bank) due to having imported money for the purchase of property; they are on the lists of the municipalities as payers of local property tax *(IBI)* and the *Hacienda* has their tax declarations in Spain (both as residents or non-residents).

A letter then went out to all who had not included the house on their German tax declarations, asking how they financed the purchase and if they had income from letting or sale. In cases where the properties are of higher value, German tax inspectors made investigations on the spot and even searched houses in company with their Spanish colleagues.

German authorities calculated that there were 400,000 German families with a property in Spain and that €8bn in unpaid taxes was collected.

10.9 Tax fraud

Tax evasion in Spain is not a criminal offence unless it exceeds €120,000 in a tax year. If so, the penalty can be one to four years imprisonment as well as a fine of up to six times the tax evaded. The offender may lose the right to receive government grants, government benefits, and tax and social security incentives, for a period of three to six years.

The penalty is likely to be in the higher of the above range where:

- False names have been used to hide the real taxpayer.
- Forged documents and fraudulent structures have been used.

Money laundering suspicions must be reported, but are restricted to handling the proceeds of crimes which are punishable by more than three years imprisonment. Thus, tax evasion of less than €120,000 is not reportable.

10.10 Clampdown on under-declarations: property sales and rentals

The Spanish tax authority's latest innovation to crack down on tax fraud is a specially designed computer program that immediately informs the tax authority, the *Agencia Tributaria,* of the details of any property sale in Spain. In particular, tax inspectors will be looking to see if the declared sale price is realistic compared to other property prices elsewhere.

The program links tax authority computers to those of public notary and property register offices throughout the country. Fraud in relation to the purchase and sale of property accounts for a fifth of the entire underground economy of the country according to some experts.

The Government is keen to completely wipe out Spain's age-old custom of declaring a lower sale price than the actual price (with the difference usually paid in cash), done to dodge capital gains and other property taxes. Buyers need to beware – when the price of a property is under-declared the seller pays less capital gains tax (CGT), but when the buyer comes to sell they will in effect have an increased CGT bill, taking on the burden of the original seller's under-declared amount.

In a bid to sweep the property sector clean, the number of inspectors focusing directly on property tax was intended to be stepped up to 40% of the tax agency's workforce, a 25% increase on the previous year. This amounts to a team of over 2,000 dedicated tax officials.

The tax agency is also intent on tightening up on the under-declaration of income earned from letting property in Spain. Six out of 10 rentals in Spain, almost a million properties, are not declared to the authorities, losing the government an estimated €1.8bn a year.

Property tax evasion is being steadily wiped out by the Spanish tax authority as new ideas are being continuously introduced to expose fraudsters and to get the message across to Spanish nationals, and the huge influx of foreigners who buy property, that Spain's tradition of tax dodging will soon be a thing of the past.

In the drive to stamp out under-declaring a property's price on a sale and purchase, laws exist forcing buyers to declare in the purchase documentation whether the property was to be purchased by mortgage, cheque or cash. Tax officials can then check sellers' and purchasers' bank accounts to verify the price paid. The law also stipulates that the owner's tax identification number should be logged on the property register so that the authorities can check if all the relevant taxes have been paid. Another ploy is to enforce the inclusion of the property registry number on the electricity bill which shows up whether property is tenanted so that undeclared rental activity can be traced.

Another campaign launched by imaginative taxmen was to plough through newspaper classified advertisements and property-related websites, scanning for rental lists as well as chatting to local business people and hoteliers to prise out names of property owners who were letting their property and who were not registered for tax.

The Tourism Council in Valencia had previously warned that it would issue fines of up to €90,000 on Costa Blanca property owners who illegally let their accommodation without their properties being correctly registered and the taxes paid.

The Valencian Government introduced new legislation to better regulate the tourist rental sector, where it was estimated that such illegal accommodation represented as much as 75% of the market for visitors to the region.

The *Guardia Civil* and the National Police initiated an operation to check all rented accommodation in the province of Alicante aimed at detecting members of criminal organisations living in rented accommodation and who are not registered with the authorities.

The police asked property owners to provide them with the identity document of the tenants.

More recent proposals have included requiring all individuals who wish to let their properties to register at the local Town Hall, and apply for a permit to let the property. All such properties would have to follow health and safety requirements, etc.

Property tax fraud is one of Spain's biggest areas of tax evasion. Property owners in Spain have to declare capital gains on sales, income tax on rental earnings, as well as succession tax on gift or inheritance of a Spanish property. In addition there are local property taxes to be paid. Unpaid tax can be a charge against the property and unless the appropriate checks are made, a new owner could find themselves with a hefty back tax bill.

Brits abroad

- There are estimated to be one million Britons living in Spain, but many are not registered with authorities.
- Britons are now the biggest group of European foreign nationals living in Spain.
- Spain has one of the fastest rising rates of immigration in Europe, with 10% of its 46m population from abroad.
- Britons are starting new businesses at a faster rate than Spaniards, according to Spanish Government figures.

10.11 Anti-avoidance rules

Spain has introduced anti-avoidance measures that affect various taxes. The most important measures may be summarised as follows:

Corporate tax

Entities situated in tax haven or zero tax territories are deemed to be tax resident in Spain where their main assets are situated in Spain or their main activity is carried on in Spain. This presumption allows for proof to the contrary.

Transfer pricing rules were changed so as to expand the concept of 'related entity', and widen the scope of transactions that are deemed to take place at market value.

Non-resident income tax

Persons or entities that are resident in countries or territories where there is no effective exchange of information with the Spanish tax authorities are obliged to appoint a tax representative in Spain if they are owners of assets situated in Spain or have rights that are complied with or executed in Spain, except quoted shares. If there is no appointed representative, the manager of the assets or rights shall be considered as such. Beforehand, these entities only had to appoint a tax representative if they engaged in certain qualifying activities or if they were requested to do so by the tax authorities.

The transfer of shares or participations in entities resident in countries or territories where there is no effective exchange of information with the Spanish tax authorities, and the assets of which consist mainly of Spanish real estate, held directly or indirectly, are now liable to CGT in Spain based on the fair market value of the Spanish real estate. The real estate assets guarantee the payment of the corresponding taxes.

Transfer tax

Transfers of shares in real estate companies (those with at least 50% of their assets invested in real estate in Spain) are subject to transfer tax when, as a result of the transfer, one person acquires the control of the company. This measure was introduced to make it more difficult for taxpayers to avoid this tax.

10.12 Priority of debts: tax vs mortgage

Under *Ley 58/2003* of the *General Tributaria* (17/12/03) the Spanish tax authorities have the right to secure tax debts against the taxpayer's assets under a process known as *procedimiento de apremio* (liability order), in the following order:

- Cash or bank deposits.
- Credits and rights exercisable immediately or with short notice.
- Salaries, wages and pensions.

- Real estate.
- Interest and any other income.

It is therefore possible for the Spanish authorities to put a charge on a property in relation to unpaid tax liabilities such as, for example, unpaid succession tax. The charge by the authorities would need to be officially registered against the property at the land registry.

However, the tax authorities do not have priority over a properly registered mortgagee or other creditor who has existing rights in relation to a property, provided that the registration of the mortgage or other charge was made before the registration made by the Spanish tax authorities *(art 77 – Ley 48/2003).*

So, if there is an existing mortgage on a Spanish property that has been duly registered against the property, then on sale of the property this takes priority over any charge subsequently placed on the property by the tax authorities.

Note, if the tax still remains unpaid at sale then the register of deeds will include a 'marginal note' to this effect, as the liability will then become the purchaser's liability (as it passes with the property) to be reclaimed against their assets in the order given above (*art 79 of Ley 58/2003* and *art 100* of the regulations for succession and gift tax).

CHAPTER 11

The Canary Islands and Taxation

11.1 Introduction

The Canary Islands is designated a Special Zone – the *Zona Especial Canaria (ZEC)* – for economic and business tax purposes. It aims to encourage the economic development of the islands and diversification of manufacturing and service sectors. The framework for the *ZEC* was approved in January 2000, and as a result of it many Spanish holding companies are incorporated and domiciled in the Canary Islands. One reason is because they do not pay the 1% capital tax payable if incorporated in mainland Spain. A holding company is very often a joint venture between two or more very large companies, and can have an issued capital running to many millions of dollars; 1% of those millions can be quite significant.

The *ZEC* was initially authorised by the European Commission to run until 31 December 2008, but this has now been extended to 31 December 2019. New companies can be registered until the end of 2013.

There is no difference to personal taxes (income and capital gains tax) in the *ZEC*.

11.2 *ZEC* tax benefits

The special tax regime *(Regimen Económico y Fiscal, REF)* operating within the *ZEC* includes freedom from VAT, despite being part of the EU. Instead there is a sales tax, the Canarian indirect general tax *(IGIC)* which operates in a similar way to VAT but at lower rates. The standard rate is 5%, and while there are higher rates of nine and 13% there is also a reduced rate of 2% and exemption for certain basic goods and services (eg, telecommunications). *ZEC* companies do not pay *IGIC* on the delivery of goods or provision of services between companies, or on imports.

Other tax benefits that *ZEC* entities enjoy include exemptions from capital transfer tax and stamp duty on the purchase of goods or rights relating to the conduct of their business activity within the geographical area of the *ZEC*, and on company operations other than liquidation; enhanced tax credits on some forms of investment; and tax relief for profits re-invested in qualifying assets within three years.

ZEC entities are resident Spanish companies, and can benefit in full from Spain's international tax treaties and the EU Parent-Subsidiary Directive.

The general rate of corporation tax payable in Spain is 30%, or 25% in the case of small and medium sized companies on the first €120,202 of profit, but *ZEC* entities can benefit from a much lower rate of 4% on profits derived from the *ZEC* activity. This fixed percentage will apply to profits up to certain limits according to the type of activity and the number of jobs created.

For example, if the company creates from three to eight jobs (or five to eight on Gran Canaria and Tenerife) the special rate of 4% can apply to taxable profits of between €1,125,000 and €1,800,000 depending on the activity. Where between nine and 12 jobs are created the range increases to between €1,500,000 and €2,400,000 of profit – and so on, up to €75m to €120m of profit with more than 100 workers. Industrial activities are favoured over service activities.

11.3 General information

The Canary Islands are an integral part of both the Kingdom of Spain and the European Union. There are seven islands; the two main islands are Tenerife and Gran Canaria, the others being El Hierro, Fuerteventura, La Gomera, Lanzarote and La Palma. Together they administratively form one of the 17 Autonomous Communities (or regions) of Spain, each with their own parliament and devolved government functions. They are located in the Atlantic Ocean off the African coast of Western Sahara about 1,050km from mainland Spain, and stretch east-west over 500km. The population is about 2.1 million, the great majority of which (about 1.75m) live on the two larger islands.

It is important to note that all European Union policies are fully applicable in the Canary Islands. Nevertheless, in recognition of the special needs arising from the geographical location, a number of highly favourable exemptions and exceptions in the application of European Community law have been established as described above.

11.4 The economy

Over 80% of the economy is based on tourism. Over nine million tourists visit the Canary Islands each year, with the result that most of the economy is based on the service industry. The *ZEC* regime has been instituted to try to change this distribution of economy to encourage a broader spread and more manufacturing. The two main cities are Las Palmas on Gran Canaria and Santa Cruz de Tenerife. There are eight airports which are, of course, necessary for the transportation of nine million tourists. Only one of the islands has two airports, which is Tenerife, and all the major European airlines operate in the Canary Islands.

11.5 The *ZEC* in more detail

The *ZEC* was initially authorised by the European Union up until 31 December 2008, but has now been extended to 31 December 2019. Companies can be registered up until the 31 December 2013, to carry out any manufacturing, commercial or service activities within the territory of the Canary Islands. *ZEC* companies may not offer services of a financial nature.

These are the industrial activities which are authorised: fishing, food, beverages, tobacco, clothing, leather goods, footwear, paper, publishing, graphic arts, the reproduction of recorded material, building materials, machinery and mechanical equipment, electrical, electronic and optical materials and equipment, chemicals, recycling, wholesale and distributive activities, packaging, bottling and canning. It is quite an extensive list. Authorised service activities are: transport-related activities, IT, natural resources, waste disposal, telecommunications, advertising, research and education training, consultancy, legal and fiscal services and others except financial services.

A *ZEC* service company may be established in any part of the territory of the Canary Islands, with no restriction, but production, processing, handling and distributive activities have to be located in certain designated areas which are near ports and airports. There are three such areas on each of the two main islands, and the five smaller islands also each have their own designated areas for industrial activities.

11.6 *ZEC* conditions

A *ZEC* entity has to satisfy certain conditions. It must be newly incorporated. At least one of the persons authorised to manage it and act on its behalf must be resident in the Canary Islands. The minimum economic investment on Gran Canaria and Tenerife is €100,000 in fixed assets relating to the activity, €50,000 on the other five islands, and this investment must be made within the first two years after registration. The company must employ at least five persons in the *ZEC* area within the first six months of authorisation and must retain this number of employees as an average throughout its lifetime. The minimum is reduced to three employees outside Gran Canaria and Tenerife. A *ZEC* company must present to the fiscal authorities a description of its activities, with particular attention to its economic viability, international focus and contribution to the economic and social development of the Canary Islands, so that the authorities can see that there is real substance and it is not just a paper company.

11.7 *ZEC* registration

The *ZEC* consortium is an administrative body that works closely with the national Government of Spain. All the laws applicable in the Canaries are exactly the same as those in force in mainland Spain; the consortium is a body which supervises the *ZEC*, and in its turn is supervised by the Spanish Treasury. It also has promotional activities and is responsible for promoting the *ZEC* worldwide.

The process of registering a *ZEC* company begins with an application and a description of the intended activities. The consortium's board controls and supervises all the documentation. It decides if the application is in conformity with the law, and it must inform the applicant of its decision within two months. Once authorised, the company can then be registered with the official *ZEC* Register *(ROEZEC)* managed by the consortium board, on presentation of the appropriate documentation.

A Spanish notary first notarises the public deed of constitution (the Articles of Association), a fiscal identification number *(CIF)* is obtained and the company registered with the tax authorities and the main companies registrar *(Registro Mercantil)* and then, with all documentation in place, finally in it can be registered in the official *ZEC* register. The *ZEC* registration fee is approximately €750; there is an annual fee of about €1,100 and the usual documentary and other fees range from €60 to €300. These are paid to the notary and the registrar; they are not high and they are the same for the incorporation of any other company.

11.8 Canary Island's letting restrictions

The rules are different depending on whether you own a tourist complex or a residential letting.

Tourist complex

If you own a touristic complex (licensed with the Department of Tourism) then there must be a sole agency who runs all the letting on the site. This can either be the community itself or an outside agency. Either way the sole agency must have at least 50% of the properties under its control in order to operate as the sole agency. The site must be licensed by the Department of Tourism and owners can only rent out through the sole agents. If there is no sole agency because they cannot achieve more than 50% of the lettings then **no one** on that site is allowed to rent.

Residential letting

If you own on a residential villa or apartment site then there are no letting restrictions on individual owners providing you agree the terms directly with your guests.

You are quite entitled to advertise and arrange all the rentals yourselves. You are acting legally providing you ensure the guests are fully aware before they arrive that it is a private apartment on a residential site with no on site reception and that you provide them with a 24-hour emergency contact telephone number of a person who will be responsible for the guests during their stay.

CHAPTER 12

Spanish Social Security and Health Care

12.1 Spanish social security contributions

Spanish social security contributions, like UK national insurance (NI), are only levied on earnings.

UK NI contributions do not give any direct entitlement to Spanish benefits or state pension, since these are based on Spanish social security contributions, although rights and entitlements to benefits in the UK based on your NI record are to some extent portable and can also affect the position in Spain in relation to health care cover, for example.

If you go from the UK to work in Spain for a UK employer, having been previously ordinarily resident in Great Britain, and expect to spend no more than two years there, you would normally continue to pay employee's Class 1 NI contributions in the UK. In some circumstances that two-year period can be extended for up to five years if it is in the best interests of the employee to continue to pay UK NI contributions for longer while working abroad. In all other circumstances, Spanish social security contributions as an employee would normally be payable.

If you work for a Spanish employer or if you are self-employed in Spain then Spanish social security contributions would be payable from the start. You personally need to affiliate to the Spanish social security system before beginning your business or starting work – although an employer has responsibility for deducting and paying an employee's contributions and will often (but not always) deal with the formalities of an employee's registration.

Social security card and registration

You will need to obtain a social security card and number in addition to your *NIE* tax identification number. The card includes the names and dates of birth of any minor children or other dependants.

There are four separate national institutions that oversee social security, social services, health and employment issues in Spain. However, as a self-employed person, you would need to register with and pay your contributions to a fifth umbrella body known as the 'the general treasury of social security' *(Tesorería General de la Seguridad Social, TGSS)* and you would pay your contributions under the 'special regime' *(regimen especial)*. The special regime also covers most company directors, employees in the home and agricultural and marine workers. The 'general regime' covers most other employees. There are 'special systems' for certain businesses including entertainment and food (catering).

Contribution calculations

An employee's contributions are in the region of 6.4%, but the calculations are complex. The employer's contributions are much higher at around 30%. Both are subject to an earnings cap of €2,732 per month. There is a minimum base, on which contributions are calculated, even if earnings are below this level.

If the salary is less than €540 per month, then minimum contributions are paid based on €540 and not the lower actual salary.

The amount of contributions for a self-employed person also varies within a range and the **minimum** amount payable is in the region of €225 per month.

12.2 Spanish health care

There is a public health system, with all members having the option of buying additional private insurance.

The state scheme covers only around 75% of the cost of treatment and the other 25% must be paid by the patient or through a supplementary 'top-up' insurance scheme. Completely free treatment is only available in a limited number of hospitals, though waiting lists tend to be very long, especially for non-emergency treatment.

To access the Spanish state health care system you must either have a Spanish social security record, or be in receipt of a UK pension and have a Form S1 (see below).

The vast majority of people in Spain are covered for health treatment under the state system. The 'National Health Service' owns 90% of Spanish medical services and equipment and runs more than 300 hospitals. However, approximately six million people (60% of the population) also take out complementary private health insurance in addition to the basic 'top-up'. This offers them a wider choice of medical practitioners and hospitals and, more important, they are not restricted to public waiting lists.

If you are not covered by Spanish social security (or are otherwise exempt – see below), you would be well advised to take out private health insurance given the average health care cost in Spain.

Cover and premiums can vary, and independent consultants such as PHA Expat (**see Appendix 1**) can provide experienced assistance.

Public health benefits include general and specialist medical care, hospitalisation, laboratory services, discounted drugs and medicines, basic dental care, maternity care, medical aids and transportation.

12.3 Spanish health care eligibility: Form S1

When you reach UK state retirement age and are in receipt of UK state retirement pension (or if you are in receipt of long term incapacity benefit/employment support allowance, severe disablement allowance or bereavement allowance) you should find that you and your dependants are permanently entitled to free basic cover under the Spanish health care system (under what is known as the Form S1 procedure).

Many were able to obtain temporary cover for health care in Spain for up to two and a half years under the temporary Form S1 system, but these were withdrawn in April 2014.

The appropriate Form S1 can be obtained from the Department for Work and Pensions, **see Appendix 1** for address details.

In Spain, if you do not have Form S1 you will be charged for treatment. If a form is pending, you may be able to get a refund, but only if you tell the Spanish sickness office before treatment that you wish to be treated under the EC rules and that the form has been applied for.

You make a claim to the *Instituto Nacional de Seguridad Social (INSS)* when you have the form.

12.4 EU health insurance card

A European Health Insurance Card (EHIC) was introduced into most EU countries on 1 June 2004. It replaced the Form E111 which was used by individuals to obtain emergency medical treatment during a temporary stay in another EU country.

The E111 has now been abolished.

The EHIC is valid in all EU countries as well as Iceland, Liechtenstein and Norway. It can be used to obtain treatment by EU individuals and their dependants travelling in another EU country.

Only state-provided treatment is covered and individuals will receive treatment on the same terms as nationals of the country visited. Private treatment is generally not covered.

Note that each family member needs to obtain a separate EHIC, including children.

12.5 Health insurance

Being ill is always a stressful and worrying experience. Being ill when you are abroad can make the experience many times worse. That is why you need a comprehensive health insurance plan you can rely on, from a health insurance provider you can trust. If you intend to live or work abroad you need to consider your health care protection even within the EU. Travel insurance is not adequate or suitable for the job as this will only repatriate you to the UK, as opposed to paying for treatment locally (other than for emergencies). So who do you choose? With so many insurers offering numerous options, it is important to obtain the right policy to suit your needs.

There are a number of well-known companies (eg, BUPA, AXA PPP) who provide insurance, but there is also an excellent range of lesser known companies who provide good insurance value. The easiest way is to use a specialist broker as they can obtain quotes from a range of companies, but will also be familiar with the policy terms (and they can vary significantly). Beware, however, of changing insurance companies as the new one may exclude any pre-existing known medical conditions whereas your current insurers may still include these. Also beware of any exclusion periods (eg, cover only starts after six months of commencement of the policy).

CHAPTER 13

Personal Financial Planning and Investment

13.1 Introduction

This chapter is intended as a general guide to formulating a sensible tax and investment strategy on leaving the UK to live in Spain. It does not go into detail about the definition of domicile or residence as these terms are fully explained elsewhere in this book. Ultimately an adviser who is familiar with both the UK and the Spanish tax rules is invaluable, as they can accurately advise you on your own particular circumstances. Remember that the rules set out in this book are not necessarily consistently applied by the tax authorities. You should always seek local advice (or an adviser who is familiar with local practice).

13.2 The facts

Most financial plans either ignore the precise circumstances and objectives of the individual, or they get it wrong. For example, a common pitfall is that many individuals living in Spain do not think that they are Spanish tax-resident because they do not really understand that staying in Spain for more than 183 days in a calendar year automatically makes them tax resident in Spain. Many British people believe that by being resident in Spain they are completely outside the UK tax net, whereas in reality they may remain liable for UK inheritance tax even if they have lived in Spain for several years. Only by a deep understanding of the facts can one even begin to establish effective tax and investment planning.

13.3 The fiscal nomad

As an individual you can be 'fiscally nomadic' meaning that you are not liable to lifetime tax anywhere in the world, save for some sort of income or gains sourced in some countries. This is because you are not resident for tax purposes in any country.

In practice, this is a position which is usually neither a practical nor permanent way of life for most people. However, it may be very advantageous to be a fiscal nomad for a short period or 'window' during which you are not tax resident anywhere; this may be a year in which sizeable capital gains are made. You have to understand the precise rules of UK residence, Spanish residence, and the tax authority's interpretation of the timing of a disposal, before you can safely step through the tax minefields and become a legal fiscal nomad.

Please note that if a UK domicile is a fiscal nomad they will retain their UK domicile status and therefore be subject to UK inheritance tax on their worldwide estate.

13.4 Assets and liabilities: income and outgoings

An accurate list of your assets and those of your spouse or partner, distinguishing those in joint as opposed to sole names, is a vital starting point. Where assets are liable to substantial taxes if disposed of, this fact must be considered before any disposals are made.

Listing your income and your outgoings is also important, along with a list of potential major one-off expenditure requirements.

13.5 Investment strategy

Before you can do this you need to work out your investment strategy, ie, how to direct your funds so as to meet your objectives.

13.6 Choosing an investment adviser

You now need to find yourself an investment adviser. If the amount which you have available for investment is less than about £20,000, this may be easier said than done.

Why use an investment adviser? After all, you have already developed your investment strategy, which is the key thing; you are reasonably intelligent; a dab hand with a pair of scissors; and if you take all the quality Sunday papers for a month, you will end up with a pile of coupons from a wide selection of investment houses covering a number of different investment structures. Well, you could do it that way – in the same way as you could buy your house overseas by walking round the town for a couple of hours looking for estate agents' signs and buying the one that looks prettiest from the outside. You will have restricted your choice to those properties that you have seen advertised; will have done no research on what else might be available; will have made your choice without sufficient information, and will have no idea if the price you are paying is realistic. No one in their right mind would do that if they were going to spend £150,000 plus on a house; so why do it when you are going to establish an investment portfolio for a similar sum?

No one wants to spend money for the sake of it. If you are going to use an investment adviser, what are you getting for the fee they will charge, or the commission that they will earn – which amounts to the same thing as, ultimately, it comes out of your pocket? What you ought to be aiming to get is:

- The benefit of their experience and expertise.
- Their knowledge of what is available in the marketplace.
- Knowledge of which companies use conservative projections and which may be over optimistic.
- Knowledge of investment managers who have performed well in the past and an assessment of whether they may have had internal changes making it less or more likely that they will continue to do so in the future.

- The ability to consider your investment strategy impartially and determine if it is really appropriate to your needs.
- The ability to select from alternative investments in order to satisfy your objectives.
- A willingness to keep your investments under review and make or advise changes where appropriate.
- A personal contact responsible, over a period of years, for your investments (not a firm with a rapid staff turnover so that you are dealing with someone new every time you phone).
- Someone who is reasonably accessible and who will provide you with a phone number where you can contact them during normal business hours in Spain.
- Someone who operates through established local offices in the place where you live, even if they are based in the UK, indicating a commitment to provide a service to clients in your area.
- Sound administrative back-up.
- Someone with a thorough knowledge of UK and Spanish tax in order to ensure you are not exposed to unnecessary tax both now and in the future.
- Someone whose choice of investment is motivated by your needs, not the commission that they can earn by introducing you to a particular product.
- Someone who is not going to run off with your money.

Before considering how you find an adviser, you need to decide where you should look. Should this be in the UK before you go, or in your new country when you get there? To an extent this depends on where you are going and for how long. If you are retiring overseas or expect to work overseas for 10 or 20 years, look at what advice is available in your new country and how closely investment advisers are regulated there by the authorities. It makes most sense to have your investment adviser in the country in which you live, other things being equal, because they can get to know you better – and thus better appreciate your real needs – than the UK-based adviser, and will be more aware of changing circumstances in your new country that are likely to affect your investments, tax position and objectives.

However, you ideally want someone with whom you can easily communicate and make contact. This may be difficult in non-English speaking countries, and yet you are not necessarily going to find a satisfactory adviser simply because they speak English. This is a lot less of a practical problem in Spain. You also want someone who is honest and reliable. While this is to an extent a matter of luck, you are less likely to engage a crook in a jurisdiction which regulates financial advisers than in one that does not.

13.7 The main types of investment

If you decide to do it yourself and do not engage an investment adviser, you will have to select your own investments. However, you need to know what types of investment vehicles are available. If you use a financial adviser, it is helpful to understand the implications of what they are investing in. It might therefore be useful to briefly describe the main options.

13.8 Direct stock exchange investments

Investing on the Stock Exchange requires not only skill but also good research, staying closely in touch with the market, and a degree of luck. If you have a minimum of £250,000 to invest in stocks and shares you could find a stockbroker to manage it for you, or ask your investment adviser to find you a stockbroker. If you are an avid reader of the *Financial Times* and the *Investors Chronicle* or a user of the internet, and like playing the market, by all means continue to do so, preferably with not more than about 20% of your savings, leaving the other 80% to be dealt with by your investment adviser. Otherwise forget stocks and shares. There is a considerable amount of paperwork with shares, dividends, scrip issues, rights issues, takeover bid papers, annual meetings, etc. Unfortunately most stockbrokers do not have any understanding of how to set up investments in a tax-efficient manner for clients living in Spain.

13.9 Unit trusts

One of the safest ways to invest in stock exchange securities is through a unit trust, offshore fund or open-ended investment company (OEIC). Each of these is a sort of investment consortium. The management company takes in several million pounds from investors in small blocks of £500, £1,000, £10,000, etc, and invests them in a pot together. Accordingly, for your £500 you get a small share in a fairly large number of investments. There is no capital gains tax within these arrangements. The dividends they receive suffer tax in theory, but in most cases tax is withheld by the company that pays the dividend and this is likely to cover the fund's tax liability on the income. If you are living in a country that charges capital gains tax or, like the UK, exempts only the first slice of gains, these arrangements can be attractive until you want to sell the units themselves.

The main snag is that these funds normally invest in a specific investment sector. Accordingly, if you want to spread your investment over different markets and sectors, which you ought to do, you need a range of funds. You then really need someone to manage these, ie, to decide when to go out of one fund and into another, whether to switch from one company's Japanese fund to an alternative, etc. Most funds have what is called a 'bid-offer spread'. This is the difference between the price at which the investment house will buy and the price at which it will sell units or shares in the fund. It is normally around 6%.

Therefore, at any time you switch into a different unit trust, you pay up to 6% of your capital, and the growth in value of your investment in the new fund has to make good this charge before you see any benefit from the switch. There are some tax planning structures in which you can hold such funds where substantial discounts are available, often reducing initial costs to between nil and 0.5% (see following section on investment bonds).

13.10 Offshore investment bonds

An offshore investment bond is a tax-efficient single premium life insurance policy. It offers a simple way to enjoy a return from a spread of investments, as the assets of the bond can be invested in a diverse range of underlying investments including international equities, government bonds, cash, or a mixture of different types of investment. The pure life insurance component is only a very small element of the bond. The investment management is carried out by investment experts of the company issuing the bond.

The cost of switching investments is far less than for direct holdings in unit trusts, but there is usually a management charge within the fund. If the insurance company is in the UK it pays tax on the income and capital gains within the funds. A number of leading insurance companies based in countries such as Luxembourg, Isle of Man, and Guernsey, issue offshore investment bonds (these are generally known as personal portfolio bonds) which are not internally taxed. These offer a choice of underlying funds which are ideal for expatriates. The choice available includes specialty funds so that you can direct your investment into, say, the Japanese or UK stockmarkets. Others offer a mixture of equities, or equity and bond investments, thus offering investors a balanced portfolio of investments. For example, one market leader offers the choice of four managed funds, each invested to reflect a different degree of risk and to meet differing investment objectives, eg, for income or growth, or a mixture of both.

You can choose to channel your investment into one or several of these underlying funds. Normally, you can also switch between the funds without cost. This type of insurance bond offers a great deal of investment choice and you are able to redirect your investments within an investment bond to take account of differing market opportunities, normally without charge. There is normally no tax consequence on the switching of investments between funds.

The investment bond issued from an EU centre can be an extremely tax-efficient investment vehicle for Spanish residents, although growth generated by a bond held in a jurisdiction blacklisted by Spain is not. A bond might be issued by an insurance company in, say, Luxembourg. Under Luxembourg legislation such investments bear no tax on the investment profits. All income and gains earned under the bond therefore accumulate tax- free. The bond is 'resident' in Luxembourg, remaining a Luxembourg asset and, therefore, your place of residence – whether it is Spain, the UK or elsewhere – does not affect this tax efficiency.

Bonds also offer easy access to investment profits. You can take regular withdrawals from the bond. This can normally be taken on a quarterly, half-yearly, or even on an annual basis. The proceeds of the bond can often be paid without deduction of any tax and there is often no automatic notification to any revenue authority. Sometimes beneficial fixed rates of tax may apply where you opt for them. The tax of withdrawals taken from the bond will, of course, depend on the rules of your adopted country. Responsibility for the declaration of withdrawals to the appropriate revenue authorities ultimately rests with the investor.

13.11 Risk analysis

One of the difficulties for the private investor is knowing how much risk he should take. This is not easy to judge. While the stockmarkets generally produce far higher returns over a five-to-ten-year period than most other investments, this is not guaranteed and such investments can, of course, go down in value.

Few investors should place all of their investments into the stockmarkets. Investment performance is likely to be far better by diversifying among guaranteed types of investments and funds, bonds providing a fixed income (although these can also go down in value), and shares. Numerous studies have shown that diversification of a portfolio in this way will produce better long-term results, with less volatility, and reduced risk.

CHAPTER 14

Taxation of Trusts in Spain

14.1 *DGT* ruling May 2010

On 14 May 2010, the Spanish tax authorities *(Dirección General de Tributación)* issued a 'formal response' to a series of questions posed in relation to the tax treatment of a Gibraltarian trust set up by a Spanish resident.

The ruling treats the trust as transparent, ie, the income and gains within the trust are treated as belonging to the settlor and are subject to income tax or capital gains tax in his hands.

The ruling makes no mention of the wealth tax treatment, although it should be assumed that the Spanish tax authorities would treat the trust assets as remaining in the settlor's personal wealth.

14.2 Impact of ruling

The Spanish authorities will rely on the ruling to treat all trusts as transparent, regardless of their jurisdiction. Under the ruling, the tax treatment of trusts in Spain is as follows:

14.3 Income/gains within a trust

The income and gains of a trust are treated as belonging to the settlor, and are taxable on an arising basis, regardless of when or whether they are distributed to a beneficiary.

14.4 Withdrawals from a trust

If made by the settlor, no tax is due, as the settlor is liable to tax on an arising basis on the income and gains arising within the trust where s/he is still alive.

Distributions to anyone other than the settlor will be treated as a succession for Spanish tax purposes, and the normal Spanish succession tax rules will apply (so a gift if the settlor is still alive, and an inheritance if the settlor is deceased).

14.5 Wealth tax

The ruling makes no mention of wealth tax, although it should be assumed that the Spanish authorities would treat the trust assets as belonging to the settlor and therefore forming part of his personal wealth.

14.6 Succession/gift tax

No succession tax is payable on the transfer of assets into the trust, since the assets in trust are still treated as part of the settlor's personal wealth. Similarly, no succession tax is payable when assets are distributed from a trust back to the settlor.

When the settlor dies, this is a taxable event for Spanish succession tax purposes and Spanish succession tax applies depending on the relationship between the settlor and the beneficiaries. The usual succession tax rules apply, ie, tax is payable where the beneficiary is resident in Spain or where the asset is located in Spain. The various reductions given in the different autonomous communities are available.

14.7 Taxation of the trustees

The overseas trustees are not liable to Spanish taxes because they are not operating in Spain.

14.8 Life insurance policies in trust

The settlor should treat the underlying asset (the PCP) as if it is his own. The following tax treatment applies:

- If the trust is set up by way of loan account, the settlor does not include the loan account on his Spanish wealth tax return (where applicable), but instead the surrender value of the PCP;
- On withdrawals from the PCP (even if not distributed), the normal Spanish income tax rules will apply as if the policy was held directly by the settlor (and if an approved policy, the provider will withhold tax in the normal way);
- The succession tax position on death is a grey area. Advice should be taken.

CHAPTER 15

Tax Planning Pre-Spanish Residence

15.1 Planning early helps

There is no question that planning early helps. You can evaluate your options, quantify your tax liabilities, plan your investment strategy, decide on ownership structures, and take steps to avoid tax on leaving the UK while minimising your taxes in Spain. You need an adviser who is knowledgeable about both UK and Spanish tax rules. Remember that the tax rules set out in this book are not necessarily consistently applied by the tax authorities and there is no substitute for experienced advice.

15.2 Not a lot of people know this

There is no law which states that you must be tax resident somewhere. It is possible not to be tax resident anywhere in the world without breaking any laws. This may be a difficult way of life, as you cannot settle in one country, but it works for some.

If you are going to live in Spain permanently, it might be sensible to have a short period of time not being tax resident anywhere. Having become a non-resident of the UK, and before becoming an official tax resident of Spain, you can dispose of assets free from UK or Spanish capital gains tax (CGT) (subject to certain conditions), and set up investment structures which will be tax-efficient in Spain once your residence begins.

To achieve this, you have to understand the precise rules of UK and Spanish tax residence and avoid a number of tax minefields.

15.3 Pensions

Planning pensions properly can save you a small fortune in Spain especially if you fit into the annuity system. You will need specialist advice on this matter.

15.4 CGT

You may be able to avoid tax in both the UK and Spain if your gain falls into any of the exemptions available, or the reliefs granted can be large enough to reduce your tax liability to an acceptable level.

15.5 Employment

It is possible to be a Spanish tax resident and avoid Spanish tax on your employment income, up to a certain level, if the employment is undertaken outside Spain in a country with a suitable double tax treaty. The 'Beckham' tax break may also save you tax if you work for a Spanish employer and have not been tax resident in Spain during the previous 10 years.

CHAPTER 16

Shedding UK Residence

16.1 Introduction

Determining your residence status is important as it affects the taxes you pay, and where you pay them. Broadly (and subject to double taxation treaties), if you are non-UK resident you will only be liable to UK tax on any income which arises within the UK but if you are UK resident you will be liable to tax on your worldwide assets.

The rules surrounding whether you are UK resident or not changed on 6 April 2013 and the UK statutory residence test came into force.

Please note that the new UK statutory residence test, designed to make determining your residence status for UK purposes easier, is highly complex (the guidance is over 100 pages long). If you are in a position where you need to determine your residence position based on the UK rules, you should always seek advice. This chapter is only intended to give an overview of the rules but it is imperative that you do not try to deduce your position based on these, as they are very simplified.

16.2 Prior to 6 April 2013

Up until 6 April 2013, you would always be treated as resident in the UK for a tax year if you fell within either of the following criteria:

- The six-month rule, ie, you were present in the UK for 183 days or more during the tax year; *or*

- Provided you had not made a 'distinct break' from the UK and you were present in the UK for an average of 91 days or more per annum measured over a period of up to four tax years.

Being present in the UK at midnight did, and still does under the new test, count as a day of presence in the UK. In practice, this means that days of arrival are counted but days of departure are discounted. If you arrive and depart the UK in the same day, this should not count as a day spent in the UK for the above residency rules, although there are anti-avoidance provisions to prevent this rule being exploited.

There is also an additional exemption for passengers in transit between two places outside the UK such that days spent in transit through the UK will not be relevant even where the individual is in the UK at midnight so long as they do not engage in any activity in the UK that is unrelated to their passage through the UK (which can include going to the theatre or having dinner with friends).

16.3 The position from 6 April 2013 onwards

From 6 April 2013, the statutory residence test is applicable. Although these new rules are more complex, they do provide more certainty in when it comes to determining whether you are UK resident or not.

You are considered resident or non-resident for the whole UK tax year, although there are certain circumstances where the year may be split into periods of residence and non-residence.

Individuals are split into two categories for the purposes of these tests:

1. A 'leaver' is an individual who has been resident in the UK for at least one of the three UK tax years preceding the UK tax year in question.
2. An 'arriver' is an individual who has not been UK resident in any of the three previous UK tax years.

The rules are slightly different for arrivers and leavers.

The statutory residence test is divided into three sets of tests (all of which include multiple tests) which you apply to your own circumstances to decide your residence status:

1. **The Automatic Overseas test** – if you pass just one of the three tests posed under this set you will be considered non-UK resident and therefore escape UK tax on income generated from your non-UK assets or earnings.
2. **The Automatic Residence test** – if you pass none of the Automatic Overseas tests, you are not automatically considered to be UK resident; instead, you look at the Automatic Residence test. If you fulfil any one of these tests, you will be classed as UK resident.
3. **The Sufficient Ties test** – if neither of the above two sets of tests provides a clear position in respect of your UK residence, the number of ties you have with the UK will determine the number of days you can spend in the UK each tax year before you will be considered UK resident. The sufficient ties test operates on a sliding scale, whereby the more ties you have with the UK the less time you can spend there without becoming UK resident. Conversely, the fewer ties you have with the UK the more days you are permitted to spend in the UK during a tax year before becoming UK resident.

The tests

You will satisfy the Automatic Overseas test, and be regarded as non-UK resident, if you pass any one of the following tests:

- If you are a leaver and you spend up to 15 days in the UK in the relevant tax year; *or*
- If you are an arriver and you spend up to 45 days in the UK in the relevant tax year; *or*
- You work full-time overseas for at least a full UK tax year.

If you do not pass any of the tests posed above, as mentioned above this does not make you automatically UK resident; instead, we must next consider the Automatic UK Residence tests. You will fulfil the Automatic UK Residence test if you:

- Spend more than 182 days in the UK in the relevant year; *or*
- Have your only home (or homes) in the UK (a 'home' does not have to be a property you personally own or rent, but can belong to a spouse); *or*
- You work full-time in the UK.

If your position is indeterminate under the above tests, you must consider your 'significant ties' with the UK. There are five tests (although only four apply to arrivers):

1. Whether your spouse or co-habiting partner and/or minor children live in the UK;
2. Whether you have 'accessible' accommodation in the UK;
3. Whether you perform 'substantive' work in the UK;
4. Whether you have spent more than 90 days in the UK in either of the two previous UK tax years; *and*
5. For leavers, whether you spend more time in the UK than in any other country. This is not a connecting factor for arrivers.

In order to determine whether you pass the Sufficient Ties test or not, you must:

- Count the number of ties you have with the UK during the tax year in question;
- Count the number of days you have spent in the UK during the tax year in question; *and*
- Using these factors consult the tables below in order to work out whether you are resident.

Days in UK	Arrivers – not resident in the UK in previous three tax years	Leavers – resident in the UK in at least one of previous three tax years
Less than 16	Always non-resident	Always non-resident
16 - 45 days	Always non-resident	Resident if at least four ties apply
46 - 90 days	Resident if at least four ties apply	Resident if at least three ties apply
91 - 120 days	Resident if at least three ties apply	Resident if at least two ties apply
121 - 182 days	Resident if at least two ties apply	Resident if at least one tie applies
183 days or more	Always resident	Always resident

The Connecting Factors

FAMILY TIE

Where you have either:

- A spouse resident in the UK (or common law or civil partner); *and/or*
- A minor child (under the age of 18) resident in the UK.

You will have a family tie.

For the purpose of the this test, HMRC will use the same principles applied to tax credits to determine if people are living together as husband and wife or civil partners (LTAHAW). To be treated as LTAHAW, the relationship should be akin to that of a married couple. Although there is no single template, factors taken into consideration are as follows:

- Whether you live in the same household;
- The stability of the relationship;

- Whether there is financial support between you;
- Whether a sexual relationship exists between you;
- Whether you share dependent children; *and*
- Whether you acknowledge yourselves publicly as such.

In cases of minors, if you see your child in person either:

- In the UK on a total of 60 days or fewer in the tax year concerned; *or*
- If the child turns 18 during that tax year, and the individual sees the child in person, in the UK on a total of 60 days or fewer in the part of the tax year before their 18th birthday, then the presence of the minor in the UK will not be seen as a family tie.

Children in full-time education in the UK are not counted provided they do not spend more than 21 days in the UK outside of term time. Half-term and other breaks during term time will be treated as such.

ACCOMMODATION TIE

This tie exists if you have 'a place to live' in the UK which is available to you for a continuous period of at least 91 days in that year, and at least one night is spent there. This would include an unlet property belonging to a spouse, civil partner or co-habiting partner. If the accommodation belongs to a close relative, the threshold is 16 days. A 'place to live' includes a holiday home and a property not actually owned by you.

A close relative for the purposes of the accommodation tie is a:

- Parent or grandparent.
- Brother or sister.
- Child or grandchild aged 18 or over.

WORK TIE

You will have a work tie if you work (for more than three hours a day) in the UK for at least 40 days (continuously or intermittently) in that year. Working includes travelling time where paid for by the employer, and job related training.

90-DAY TIE

This test is fulfilled for a year if the individual spends more than 90 days in the UK in either the year preceding the current tax year and/or the year before that one.

COUNTRY TIE

The country tie only applies to leavers. It does not apply to arrivers.

An individual has a country tie if the UK is the country in which they meet the midnight test for the greatest number of days in that year. The 'midnight test' is met if the individual is present in that country at the end of the day.

Care must be taken to ensure that days spent in the UK are properly recorded (indeed, it is sensible to ensure that you keep detailed records of where you have spent your time, particularly if you travel a lot between countries) and that any ties that you may have with the UK are reviewed regularly to ensure that the best tax position for you may be maintained going forward.

16.4 Residence overseas

The fact that you may claim to be resident in Spain does not therefore mean that you cannot also be resident in the UK at the same time. Spain has a double tax treaty with the UK which deals with situations where an individual falls into the residency criteria of two countries at the same time, and this is explained in more detail in Chapter 3. Remember that in such a case you may be called upon by HM Revenue and Customs (HMRC) to prove that you are a registered tax resident of Spain.

16.5 Married people

The residence status of married people (or civil partners) is determined separately in the UK; it does not have to be the same for both spouses.

16.6 Immigration status, home office status, British citizenship and domicile, etc

None of these factors has any relevance in determining your UK tax residence status. They are ignored, except in unusual circumstances under the UK/Spain double tax treaty.

16.7 The date of non-residence

Technically, your UK residence extends for the entire tax year. The UK tax year runs from 6 April to 5 April each year. However, from 6 April 2013, the legislation allows a tax year to be split into two parts in eight specific circumstances.

The effect of split-year treatment is that you can qualify for non-UK resident status for part of the tax year and UK resident status for the remainder, depending on your specific circumstances. This can be useful as many UK taxes are only levied on UK residents.

The split-year rules are a complex part of the UK statutory residence test and again, it is important that you seek advice if you think that they may apply to you.

16.8 How do the authorities know where I am resident?

We are often asked how tax authorities around the world can find out whether or not you are resident in their country, and how they can then tax you. There are many points to be aware of.

In most countries it is your responsibility to make yourself known to the tax authorities if you are tax resident. If you are caught not declaring your tax residence, you can be fined, or even jailed.

There is usually a huge amount of information for instance, which is automatically passed to your new country's tax authority. This might include yacht registration, becoming a company director, buying property or receiving bank interest. The UK/Spain double tax treaty enables information to be passed to the other country, but not on an automatic basis. It is no good telling the tax authorities in the UK that you are resident in Spain unless it is true. You will immediately be asked for your tax identification number in Spain so that contact can be made to check your story (and vice versa).

As you wonder how they can tell where you have been for the last 183 days (or however many it is) as your passport is never stamped, think about the massive trail of paperwork which you leave behind you: telephone bills, mobile phone records, electricity bills, bank statements, credit cards, parking fines, correspondence with professional advisers or files with doctors/dentists, etc. All stand by to give evidence of your whereabouts on a daily basis. Several recent UK residence cases have used these paper trails to help determine an individual's residence. Where a lot of financial activity was taking place in the UK, in the absence of proof from those individuals about where they were (in many cases, they could not remember, as they had not kept records), the presumption was that they had been present in the UK on those dates.

Airline manifests are not sacrosanct. Companies can be compelled to disclose to the tax authorities the names of all holders of certain assets with Spanish addresses (the French requested details of all Gold Card holders with French addresses from American Express some years ago, for example). Spain have co-operated with German tax authorities to provide details to Germany of all German residents with Spanish property. Most tax authorities have the power to obtain this kind of information direct from third parties. Computers are phenomenal at storing and retrieving information easily.

Try firing your gardener, or divorcing your spouse, upsetting a neighbour, or falling out with a business partner. These people are well known for informing on 'ghosts' (individuals who are tax resident in a country but never declare themselves).

Most reasonably sophisticated tax authorities have the right to interview you, and possibly your spouse. Quite often the onus of proof can be placed on the taxpayer, rather than the tax inspector.

CHAPTER 17

Avoiding UK Capital Gains Tax (CGT)

17.1 Introduction

There are two ways to avoid UK CGT:

1. Becoming a non-UK resident.

2. Claiming relief as a UK resident.

At the time of writing, the UK Chancellor of the Exchequer, George Osborne, has recently announced as part of the UK autumn statement a change in UK law, to be introduced in 2015, which means that the tax planning opportunity to escape UK capital gains tax on UK residential property for non-residents has now been limited. There have already been restrictions enacted for non-resident companies owning UK real estate.

Certain other gains, such as gains arising on sales of shares realised while non-UK resident, remain exempt from UK capital gains tax in the hands of long-term non-UK residents.

As such, the information given below at 17.2, reflects the law at the time of writing but will not reflect the law from 6 April 2015 onwards in respect of UK property. At the time of writing, consultation has begun in the UK on how the law will apply, but as yet we have no firm details.

Where you have been resident for four out of the seven of the tax years preceding your departure AND you return to the UK within five full tax years of the date of your departure these rules will apply.

In these circumstances amounts received during the time away from the UK will be charged to the individual in the tax year of his return. If, however you remain non-resident for five complete and consecutive UK tax years, any of these amounts received during your absence from the UK will be exempt from UK capital gains tax. Please note, however, that disposal of UK assets as Spanish residents may give rise to Spanish capital gains tax.

17.2 Current position for UK property

Currently, where a UK resident moves abroad and at the date of leaving the UK owns UK property, as long as the individual remains outside of the UK for at least five complete and consecutive UK tax years, any property disposed of during his absence will escape UK capital gains tax.

17.3 UK real estate from 6 April 2015

As mentioned above, from April 2015, non-UK residents will be liable to capital gains tax on gains arising on disposals of UK residential property.

The information currently available as to how this new measure will operate is incomplete and full details will be published as part of the UK Budget 2014. So far only 'future gains' have been referred to when discussing this new measure, which indicates that only gains arising from 6 April 2015 will be subject to tax; however it is not currently clear whether this will be the case, or whether the full gain since acquisition will be taxed.

In any case, from 6 April 2015 where an individual sells a UK property while they are abroad, regardless of whether they return to the UK or not within five years of leaving, a least some part of any gain arising on disposal of that property will be subject to capital gains tax.

Many individuals may opt to dispose of a UK residential property before the measure comes into force. However, before selling in reaction to these proposals, consideration should be given to the tax position in their country of residence.

At the time of writing, Blevins Franks advises individuals affected by this measure to consider moving funds from UK property to tax-efficient investments within the next 15 months, although we also recommend waiting to ensure that the exact form the measures will take are fully understood.

17.4 Becoming non-UK resident

This section summarises the UK capital gains tax position for someone going to live anywhere outside the UK.

Resident in the UK means that UK capital gains tax is due on all assets throughout the world. Not resident means that you are not liable for capital gains tax even if the assets are located in the UK (unless the assets are used for a trade in the UK).

Prior to 6 April 2013, an individual leaving the UK is treated for capital gains tax purposes as non-resident the day after they have left the UK, provided they were not resident and not-ordinarily resident in the UK for the whole of at least four of the seven tax years immediately preceding the UK tax year of departure.

From 6 April 2013, the statutory residence test will apply to determine your residence position for capital gains tax as well as income tax.

The effective date a capital gain is realised is on the date the contract is made (which in some cases can even be before a formal unconditional exchange of contracts), not on completion. For example, in most property sales the contract is entered into a month before completion. Furthermore, a contract is not necessarily a document headed 'contract'. It is an agreement to sell. If a verbal agreement is entered into before the date of the written contract, it will be the date of the prior agreement that matters. If you were a UK resident at that time, you are caught.

The agreement does not have to be contained in a single document. Indeed, legally, a contract requires one party to make an offer (to buy or sell an asset) and the other party to accept it. An offer and acceptance are frequently in separate documents, eg, the offer may be in a letter from your purchaser's solicitor to yours, and acceptance in a letter from you to your solicitor that is faxed to the other solicitor.

'Arrangements' made while in the UK to later sell an asset while non-resident (even though they are not legally enforceable) have been held to constitute the effective date of disposal. Remember also that HMRC has the right to see all correspondence, memos, etc, leading up to a contract, and these are often requested.

If you realise a capital gain in the tax year you leave the UK, the gain is definitely liable to UK capital gains tax, subject to the annual exemptions and other exemptions mentioned in this chapter. This is also the case where you dispose of an asset in the UK tax year you leave, but after departure, unless you were not previously resident for four years or more out of the past seven years, as explained above.

To avoid UK capital gains tax, you therefore generally have to be non-resident for the whole UK tax year. Resuming residence in the UK within five years of leaving will normally forfeit the exemption from capital gains tax in relation to assets owned at departure, and any gains made in the intervening years will become chargeable in the year you return to the UK. To avoid this, you need to be non-resident for five complete UK tax years.

You need to take particular care if you are selling your family company and emigrating to enjoy the proceeds without giving the tax authorities their share. If you remain in the UK until the negotiations are at an advanced stage, you risk unwittingly entering into a contract in advance of the solicitor drawing up the formal sale agreement.

Furthermore, if you leave in the middle of the negotiations, you risk them falling through, without you there to smooth over relationships and make compromises. Of course, you can be on the other end of a telephone, fax machine and computer in your Spanish (or other) villa.

Options

A possible way round this dilemma is not to sell the business but to grant the purchaser an option to buy it which would not be exercisable until the tax year after you have emigrated. You can negotiate all the terms of the sale 'contract', as you are not actually entering into that contract. You are entering into an agreement under which the purchaser is entitled, once you have emigrated, to call on you to sell them the shares in your company in accordance with the terms of the 'contract' that you attach to the option. An option that is subsequently exercised is not a contract at the time the option is granted but becomes a contract when it is exercised, ie, when the purchaser formally requires you to sell them your shares in accordance with the option. Accordingly, your disposal does not take place until that date as far as the tax authorities are concerned. There are, however, a number of snags.

- You still have to take care to ensure that you do not enter into an informal contract to sell the shares prior to signing the option agreement. If you do, the option will be ineffective.
- In many cases such an arrangement is impracticable. During the period between the grant of the option and its exercise, who will run the business? You could stay in the UK until 5 April, but need to be overseas before the start of the UK tax year in which the option is exercised. The purchaser may not want to run the business before taking ownership of the shares. In any case, you probably do not want the purchaser meeting your customers and suppliers and getting to know how your business operates, months before you have been paid. They may decide with such knowledge that they do not need your company, and decide to set up in competition.
- The purchaser could decide in the interim not to buy your company after all. You can guard against this by the use of cross-options, ie, the purchaser would have the right to call on you to sell your shares, and you would have the right to require them to buy. However, this has its own problems. HMRC has been known to contend that if both parties are able to enforce a sale at the

same time, these rights are equivalent to the option agreement constituting a contract for sale, which is what you are seeking to avoid. Even if you ensure that the options are exercisable at different times, what is your right really worth? How can you force someone to pay you money if they decide not to honour their obligation? You can always sue. But if the case is defended you may well have to come to the UK to give evidence.

- You may lose the sale. If your prospective purchaser has lined up finance from a bank or other lender, they may be unwilling to commit to that money being available in six months' time or whenever the option becomes exercisable. If your prospective purchaser has 'psyched themselves up' to buy your business, they may lose interest very rapidly once you say that you would like to negotiate the terms but, having done that, they cannot actually buy the business for another six months. If the purchaser has lined up finance they may be committed to paying interest on a heavy borrowing for the next six months and having to leave the funds on bank deposit until the option is exercised. Guess who will be expected to bear the difference between the two interest rates?
- Finally, the UK has now enacted General Anti-Abuse provisions (the GAAR) which can catch such arrangements if there is any belief that this is not an option, but effectively a contract.

Share Exchanges

Another solution that is often suggested is to sell your shares not for cash, but for loan stock issued by the purchaser. This can only work if the purchaser is a company. There is an exemption from capital gains tax where one company acquires another in exchange for shares or securities (such as loan stock) issued by the purchaser company. The transaction is not a disposal for capital gains tax purposes; instead, the law pretends that your new shares or securities are the same assets as your original shares in your family company.

The exemption applies only if both:

- The share exchange (or exchange of shares for loan stock) is effected for bona fide commercial reasons; *and*

- It does not form part of a scheme or arrangements of which one of the main purposes is avoidance of liability to capital gains tax.

The courts have given a very wide meaning to the word 'arrangements'. If you ask the purchaser for loan stock rather than cash, or if they suggest to you that you may prefer loan stock as they know you intend to emigrate, you probably have 'made an arrangement', one of the main purposes of which is to avoid capital gains tax.

There is a clearance procedure. Your company (or the purchaser company) can tell HMRC before you enter into the sale contract that you are selling the business in exchange for loan stock and ask for confirmation of acceptance that the two conditions above are satisfied. If such confirmation is given it cannot later be claimed that the transaction constituted a disposal of your shares, provided that the clearance application 'fully and accurately' disclosed all facts and considerations material for HMRC to make a decision. If it did not, the clearance is worthless. You must form your own opinion on whether HMRC is likely to accept that a sale in exchange for loan stock shortly before you emigrate, and thus escape the UK tax net, did not form part of an arrangement designed to avoid capital gains tax. There are also anti-avoidance provisions which may apply to bring the gain back into account in certain circumstances, even if the above conditions are satisfied.

You will have to wait for your money. HMRC will, in practice, not give clearance if the loan stock has a life of less than six months and normally a longer period. While you are waiting for repayment, you are in the position where you no longer own your company but have in its place a piece of paper, which is usually unsaleable, issued by a company over which you have no control. In the 1980s many people sold their family businesses to a publicly quoted company, British and Commonwealth Holdings Ltd, in return for loan stock. It subsequently went into liquidation and the loan stock proved worthless. If the vendors had not sought to avoid the tax, they would at least have ended up with 60% of the value of their businesses, instead of losing the entire amount. In most cases, with a little pressure (and probably a small reduction in the money you receive, to cover the bank's fee) it is possible for the purchaser's bank to guarantee the loan stock. This may be a precaution that is well worth taking even if you end up paying the bank's guarantee fee yourself.

A trap: selling a business

Although one normally avoids capital gains tax by ceasing to be both resident and ordinarily resident in the UK prior to the sale, there is an important exception. A non-UK resident is liable for capital gains tax on UK assets used for the purpose of a trade carried on in this country. Accordingly, if you are self-employed or a partner, you cannot go overseas, then sell the continuing business, and escape capital gains tax. The money you receive will be for the disposal of UK trading assets, such as goodwill, and thus within the scope of capital gains tax.

Example

Mr and Mrs Black owned a successful freehold hotel which they operated as a partnership. They left the UK in February 2004 leaving a manager in charge of the hotel. In May 2004 the manager agreed to buy the hotel, giving Mr and Mrs Black a capital gain of £430,000.

Unfortunately, Mr and Mrs Black remained liable for UK capital gains tax, even though they were both not resident and not ordinarily resident.

What if Mr and Mrs Black had gone abroad, ceased the trade, and then sold the hotel as a property disposal? This would not have helped. The cessation of trading by a non-resident attracts capital gains tax as if the assets had been sold at their market value at that time.

What if the trade had not been running a hotel but a hotel consultancy so that Mr and Mrs Black could have gone overseas and taken their goodwill with them? It would then cease to be a UK asset. The Government has thought of that too. Taking a trade asset out of the UK while the trade continues is also treated as giving rise to a disposal for capital gains tax.

You should note that capital gains tax is chargeable under this rule in respect of assets of both a trade and a profession.

It is also worth pointing out that letting a UK property held as an investment is not a trade, and is therefore not liable to be caught in this trap. Even disposals of short-term holiday lettings (regarded as a trade in some sections of the Taxes Act) are not caught by this rule.

Q Is there anything that Mr and Mrs Black could have done?

A With advance planning they could have avoided UK capital gains tax by one of two methods.

1. **Incorporation** – If the business had been incorporated while Mr and Mrs Black were still UK residents, the property could have been transferred to the new company or left in their own name. The sale of the shares and/or the hotel freehold would have been free of UK capital gains tax if sold in the tax year after leaving the UK, provided that Mr and Mrs Black ultimately remained both non-resident and not ordinarily resident in the UK for at least five tax years.

   They could have put the property into the company with the business and sold the shares of the company.

   Alternatively, they could have let the property to the company (at a market rent) and sold both the shares and the property. Once let, the property would cease to be an asset used in a trade and would have become an investment property.

   It is also important to incorporate in such a way that the incorporation itself does not give rise to capital gains tax. There are several ways that this can be done. Professional advice needs to be sought to determine which is the most appropriate in the specific circumstances of the case. It is very important that the incorporation occurs before Mr and Mrs Black leave the UK.

2. **Renting the freehold** – An alternative way would have been for Mr and Mrs Black to let the hotel to the manager and no longer be involved in the business themselves. The manager signs a lease, paying a market rate. Later, having left the UK and established being non-resident and not ordinarily resident, they could have sold the rental property to the manager free of UK capital gains tax. Again, it is essential that the lease is granted before leaving the UK.

There is one word of caution about such tax planning. There have been several tax cases (in particular, one known as *Furniss vs Dawson*) which in certain circumstances enable HMRC to ignore steps in a series of transactions which have no real commercial purpose, other than the avoidance of tax.

British Forces and Crown servants overseas

If you work overseas in HM Armed Forces or in some other capacity as a Crown servant, your employment earnings are treated as if you remained in the UK, as you are considered not resident but ordinarily resident. PAYE would therefore continue to apply. However, the normal rules of residence still apply and thus, there are opportunities to save capital gains tax, income tax and inheritance tax.

17.5 Remaining UK resident

Introduction

UK capital gains tax rules underwent major reform on 6 April 2008, and since then all gains are taxed at a flat rate of 18% on income which falls within the basic rate band (£32,010 in 2013/14 and £31,865 in 2014/15) and 28% on income which falls in excess of the basic rate band, regardless of how long the asset has been owned, and regardless of whether the asset was used in a business or not. If the asset was purchased prior to 31 March 1982, the original cost of the asset is ignored and the market value at 31 March 1982 substituted, assuming that the cost is lower than the value at 31 March 1982. A chargeable asset includes all assets unless they are specifically exempt (exempt assets are listed below).

Reliefs for inflation (indexation allowance) and taper relief were also abolished, although the annual exemption (£10,900 per person for 2013/14 and £11,000 in 2014/2015) remains. In addition, a new relief has been introduced to reduce the tax payable on gains of up to £10m in certain situations, and these are discussed below.

Any losses which arise on the disposal of an asset are offset against chargeable gains made in that same tax year. If the losses exceed the gains, the unrelieved loss may be carried forward indefinitely against any future capital gains. The unrelieved losses cannot be carried back, nor can they normally be offset against any income liable to tax; they can only be offset against capital gains.

Calculating the gain

The gain is calculated as the sale proceeds, less the cost of acquiring the asset. In working out the costs of acquisition and disposal, you can also deduct incidental costs of both, such as legal or other fees, local taxes on property, loan arrangement fees, etc.

Husband and wife

A husband's and wife's (or civil partner's) residence status are determined independently, and they are taxed as single persons even if one is resident in the UK and the other is not. If they are still living together as a couple (even though one may be tax-resident in the UK, and the other non-resident), transfers from one spouse to the other are free of capital gains tax. In a famous case *(Gubay vs Kington)*, the tax authority tried to argue that the normal exemption on disposal of assets between two spouses did not apply where one was not resident and the other was resident. The House of Lords ruled against HMRC. Thus, it appears that if a husband transfers all his assets to his non-resident wife, and she sells them free of capital gains tax, the tax can be avoided (and this should also apply to transfers of assets between civil partners). Professional advice must be sought in this area and, of course, you would need to consider any capital gains tax liability in your wife's country of residence.

Entrepreneur's relief

Entrepreneur's relief is available for gains arising on disposals of all or part of a trading business (including professions and vocations, but not including a property letting business other than furnished holiday lettings), regardless of whether trading as a sole-trader or partnership, or on cessation of such a business, and of certain shares in trading companies, provided that these assets have been held for more than one year. The age of the vendor is irrelevant.

This relief effectively reduces the tax rate on the first £10m of gains to just 10% (as opposed to 18% or 28 %). This is a lifetime allowance, not just a 'one-off' relief, and therefore, if the gains on the first disposal do not exceed this amount, any balance of the allowance remaining will be available to set against future qualifying gains.

Gains arising on disposals of shares held for at least one year are also covered by this exemption, provided the individual selling the shares:

- Has been an officer or employee of the company, or of a company in the same group of companies; *and*
- Owns at least 5% of the ordinary share capital of the company and that holding enables the individual to exercise at least 5% of the voting rights in that company.

Where an individual sells shares on which relief is available, if they also sell assets belonging to them personally that are used in the company (eg, the property from which the business trades) at the same time, the relief will also be extended to such assets.

Assets exempt from UK capital gains tax (CGT)

The following assets are not liable to CGT:

- An individual's only or main private residence. There are special rules regarding this.
- One other residence occupied rent-free and without other consideration by a dependent relative, where occupied prior to 5 April 1988.

- Chattels which are wasting assets, unless used in a business. A wasting asset is one which has a predictable life of less than 50 years. Chattels are tangible moveable property such as coins, furniture, works of art, fur coats, yachts, greyhounds, etc. Currency is not exempt unless it is Sterling or foreign currency for personal use while abroad.
- Non-wasting and business chattels where the disposal proceeds of an individual item do not exceed £6,000. Items comprising a set or collection are treated as individual assets unless they are sold to the same or connected persons, in which case, the sale is treated as one sale arising on the occasion of the last sale.
- Gilt-edged securities, local authority bonds and qualifying corporate bonds. A qualifying corporate bond is a bond issued by a UK-listed company (or one traded on the USM) denominated in Sterling.
- Save As You Earn contracts, savings certificates and premium bonds.
- Prize and betting winnings.
- Private motor cars including veteran and vintage cars.
- Sterling currency and foreign currency for an individual's own spending and maintenance of assets abroad.
- Decorations of valour, if disposed of by the original holder or legatees but not by a purchaser.
- Compensation or damages for personal or professional negligence or injury.
- Life assurance policies, but only in the name of the original owner or beneficiaries, or a person who received the policy as a gift from such a person.
- Gifts of assets that are considered by the Treasury to be of national, historic or scientific interest, but breach of any conditions imposed will nullify the CGT exemption.
- Gifts to charities.
- Enterprise Investment Scheme investments that have been owned

for three years or more, where income tax relief has been given under the scheme and not withdrawn.

- Individual Savings Accounts (ISAs, TOISAs and, formerly, PEPs).

The annual exemption

Each individual has an annual capital gains tax exemption (£10,900 in the 2013/14 UK tax year and £11,000 in 2014/15). A loss made in a tax year must be offset against gains of that same year prior to applying the annual exemption but losses brought forward need not be set against gains covered by the exemption, thus preserving such losses for later years. Any unused exemption cannot be carried forward or transferred between spouses or civil partners.

Foreign currency – a note

Exchange gains on foreign currency are liable to UK capital gains tax except for currency for your own day-to-day use. This is so even if the currency is used to acquire other foreign investments or to buy a foreign property without any intervening conversion into Sterling. However, you may treat transfers between foreign currency bank accounts containing a particular currency as one account, so that capital gains tax does not then arise (see HM Revenue and Customs Statement of Practice SP10/84).

The tax treatment of foreign currency accounts can be a trap for the unwary.

Payment dates of UK capital gains tax

The normal date of payment is 31 January after the end of the tax year in which the gain was made, or 30 days after the issue of the assessment, if later. Penalties can be charged if the capital gain is not reported to HMRC before 31 January (or in some cases by 6 October in the year after the one in which the gain has been made if a tax return has not been issued). Interest can be charged on late-paid tax.

The tax can be paid by instalments, where the proceeds of the sale are received by instalments over 18 months or more, and paying the tax in one sum would cause hardship, but interest continues to run from the due date of payment (had instalments not been agreed) until the tax is paid in full.

Capital gains tax on death

No capital gains tax charge arises on death – is this the ultimate tax plan? HMRC appears to have accepted the fact that death is not a tax planning 'technique'.

Hold-over relief and gifts

Hold-over relief avoids capital gains tax on gifts of business assets or shares in a 'family' company. Gifts between husband and wife or civil partners who are living together cannot give rise to a capital gains tax liability.

Usually any other gifts (including a sale at a low price to a relative or unmarried and non-civil partner) will be liable for capital gains tax as if they had been sold, and HMRC will substitute market value for the actual sale price. If the donor of business assets is an individual or trust, and the donee is an individual or trust who would be liable for capital gains on the subsequent disposal of the asset, by making an election in writing, no capital gains tax would be due. The election should be jointly made by donor and donee, unless the donee is a trust, in which case, only the donor need claim.

If the gift attracts inheritance tax which is paid by the donee, the inheritance tax is added to the donee's base cost for future capital gains tax computations.

The hold-over relief can also be used for gifts of business assets from an individual to a company (but not vice versa). The claim can only be made if the gift is an asset used in the business.

Reinvestment relief for Enterprise Investment Scheme (EIS) shares

This relief is not appropriate in many cases for someone wishing to relocate permanently to Spain as the gains on the shares themselves will not be exempt from capital gains tax in Spain. Also, if you defer a gain into EIS shares and you become non-resident within three years of making the investment, the deferred gain will crystallise in the UK on departure from the UK.

A chargeable gain of any description realised by an individual from the disposal of an asset can be deferred by subscribing for qualifying shares as defined for the EIS. An unlimited amount can be deferred for capital gains tax purposes (even though there is a limit for income tax relief). There are specific criteria that need to be fulfilled for the relief to be available.

On the occurrence of a chargeable event, the investor is treated at that time as having realised a chargeable gain equal to the amount of the gain originally deferred. A gain triggered in this manner can itself be deferred again by reinvesting in further EIS.

It should be noted that CGT hold-over relief cannot be claimed in respect of any deferred gain triggered by a disposal of the EIS shares for example, where such shares are given to the trustees of a discretionary trust.

Also, the gain cannot be deferred by reinvesting in a company in respect of which the gain arose.

CHAPTER 18

Your UK Home while Living in Spain

18.1 Introduction

There are three aspects of continuing to own a UK home while resident in Spain:

1. If rented, what is the rental income position with regard to tax in the UK and Spain?
2. If later sold, what is the capital gains tax position in the UK and Spain?
3. UK and Spanish inheritance tax on death or gift.

18.2 Rental income: UK and Spanish tax

You will be liable for tax on rental income in the UK where the property is situated, and it will also need to be reported on your Spanish tax return. The amount calculated under Spanish rules is added to the Spanish tax calculation. You then receive a credit for the UK tax paid.

Letting your home: tax deductible costs

The following costs can usually be deducted against the rental income for UK tax purposes:

- General and water rates where paid by the landlord.
- Agent's fees.
- Legal fees relating to the letting.
- Repairs and redecoration.
- Postage and telephone costs directly relating to the letting.
- Wear and tear (see below).
- Other services (eg, electricity, gas, television licence, paid by the landlord).
- Insurance.
- Inventory fees.
- Ground rent.
- Valuation fees for insurance purposes.
- Gardening costs or window cleaning, if imposed on the landlord by the lease.
- Accountancy fees.
- Interest costs (see below).
- VAT charged.
- Council tax to the extent borne by the landlord.

Accounts

Profits or losses are calculated in accordance with normal accounting procedures, matching income and expenditure in the chargeable period, and making provision for accruals and pre- payments. Rents received in advance must be apportioned to the relevant tax year. Accounts should be prepared for each tax year ending on 5 April.

The income from all properties you own is to be pooled and treated as a single business activity, although it will still technically be regarded as investment income.

Wear and tear

If the property is let furnished, an allowance for wear and tear on furniture, carpets, fittings, etc, may be claimed. There are two methods of allowance commonly approved by inspectors:

1. 10% of the net rental income (ie, gross rents, less rates and annual payments, eg, ground rent) as a wear and tear allowance; *or*
2. The net cost of replacements, but nothing for initial purchases Once you have chosen one of the above methods, you must stick with it.

Interest

Interest on a loan is allowed as a deduction in calculating the profits of the business. A loan used in the business of property letting will generally qualify for relief whether used for the purchase of the property or to pay expenses.

The interest on an overdraft can also qualify for tax relief if it is clear that the overdraft is for the purpose of the business. However, the interest paid will not qualify for tax relief if the loan has been used to finance personal transactions. HM Revenue and Customs (HMRC) might ask for a balance sheet to be prepared for the business to prove that the owners have not become overdrawn personally.

Most people find that they have to spend money on redecoration and possibly on buying some new furniture in order to let the house. Interest on a loan to meet these costs is deductible.

Interest is tax-deductible but there is a catch if you borrow from overseas. You still have to deduct UK income tax at the basic rate if it is a non-UK bank and where the interest is derived from a 'UK source'. The term 'UK source', according to HMRC, takes into account the following factors:

- The residence of the debtor, ie, the place in which the debt can be enforced.
- The source from which the interest is paid.

- Where the interest is paid.
- The nature and location of the security of the debt.

If all these are located in the UK, then UK tax should be deducted from interest payments. This would probably be avoided if the interest was paid from funds other than the rent, and the payment was made overseas.

HMRC's Personal Tax International section has, however, indicated that deduction of tax is not required if the taxpayer is non-UK resident.

Sheltering the rental income from UK tax

There is no limit to interest deductibility on loans used to purchase or improve a property which is commercially let.

> EXAMPLE
>
> If you were intending to invest, say, £250,000 of your own money and take out a mortgage of £50,000, you could consider taking out a mortgage of around 70% (£200,000), using only £100,000 of your money rather than £250,000. The remaining £150,000 can be invested tax-efficiently.
>
> As long as your loans meet certain conditions, you can now:
>
> - Deduct interest on £200,000 of loans against rental income, not £50,000. This will reduce your UK tax bill.
> - Invest £150,000 overseas, with the opportunity of investing it in a tax-sheltered form.

Appointing an accountant in the UK

It is well worth appointing an accountant who is experienced in lettings and capital gains tax to ensure that you minimise your tax liability. In addition, the accountant can take the responsibility of dealing with HMRC and obtaining an exemption from deduction of tax by your agent. The accountant can also receive all tax assessments directly from your inspector of taxes, ensure appropriate appeals are made within the 30-day limit (which can be difficult to meet from overseas) and can prepare your annual UK tax return.

18.3 Receiving UK rental income without tax deduction

The Non-Residents Landlords Scheme applies to all rents and other income of a UK rental business paid to non-resident landlords.

How does the scheme work?

Unless you have HMRC approval not to do so, letting agents, or where there is no letting agent, the tenant must:

- Deduct basic rate tax from the landlord's UK rental income; *and*
- Pay the tax to HMRC.

They must do this for each quarter, that is for each three month period ending on 30 June, 30 September, 31 December and 31 March.

A letting agent is not necessarily a commercial agent, but essentially any UK resident who has the power to receive the rental income on behalf of the landlord as an intermediary.

Letting agents and tenants do not have to deduct tax from the rental income of a non-resident landlord if HMRC has told them in writing that the landlord is approved to receive the rental income with no tax deducted.

Any tax that is to be deducted is calculated at the basic rate on the gross rents received, less any allowable expenses paid on behalf of the landlord in the quarter.

Non-resident landlords can offset the tax deducted from their UK rental income against their UK tax bill when they complete their UK tax return and claim repayment of any excess tax deducted. Non-resident landlords can apply to HMRC Personal Tax International for approval to receive their rental income with no tax deducted.

How can I apply for approval to receive income with no tax deducted?

If you want to apply to receive your UK rental income with no tax deducted, you can get an application form from HMRC Personal Tax International. There are separate application forms for individuals, trustees and companies. You can apply on the basis that either:

- Your UK tax affairs are up-to-date; *or*
- You have never had any UK obligations; *or*
- You do not expect to be liable for UK tax.

In each case, you must undertake to comply with all your tax obligations. In particular this means that you must complete the tax returns that HMRC send you and you must pay your tax on time.

What happens to tax which is deducted from my income?

You can offset it against your UK tax liability. HMRC will send you a tax return to complete.

The tax deducted is unlikely to match exactly your tax liability. If the tax deducted is less than your tax liability, you must pay the excess. If the tax deducted is greater than your tax liability, you can claim repayment of the excess from HMRC.

The UK tax liability paid can generally be credited against any equivalent income tax liability in the country of residence, although any excess over the amount due in the foreign country will not normally be refunded.

How can I tell how much tax has been deducted from my rental income?

Your agent or tenant must give you a certificate by 5 July each year telling you the amount of tax deducted from your rental income in the year ended 31 March.

18.4 Capital gains tax: UK and Spain

UK capital gains tax: your own home

Up until recently, if you sell the property in a tax year in which you are non-UK resident and which forms part of a longer period of at least five complete and consecutive UK tax years of non-UK residence, there would be no UK capital gains tax due. However, changes to UK tax legislation mean that from April 2015 onwards, this will no longer be the case. If you think that these changes affect you, for example if you are living in or moving to Spain from the UK while at the same time retaining UK property, you should seek advice.

Provided a house has been your only or main residence at some time, the gain treated as accruing over the last 36 months (18 months from April 2014) of ownership is always exempt from capital gains tax.

There are many other rules about avoiding capital gains tax on a principal private residence (your own home), which are rather complex but worth setting out in detail as this is often a key point for the working expatriate.

MORE THAN ONE RESIDENCE

It must be borne in mind that you can have two residences available to you without owning them both. A home which you rent counts as a residence. An overseas flat or house, whether owned or rented, is also a residence.

If you own (or have available) more than one residence, you can at present elect which property should be regarded as the principal private residence for capital gains tax exemption, although this may be withdrawn or modified under the changes to capital gains tax for non-residents mentioned above. Such an election, which must be in writing, cannot apply for a period starting earlier than two years prior to the election. The election can be varied at any time by giving a fresh notice but again, the variation cannot apply to any period over two years before the date of the new election. If no election is made, the inspector of taxes will choose, on the basis of what appears on examination of the facts, which is to be the main residence at the time. A second property which you let to bona fide tenants (or keep available for letting) is not treated as being available to you.

TIME LIMIT FOR ELECTION

The time limit for making the election is two years from the date the second residence first becomes available. HMRC has stated that where the taxpayer's interest in one of the properties has 'no more than a negligible capital value in the open market' and the taxpayer was unaware that an election could be made, it will extend the time limit to a reasonable time after the individual first becomes aware of being entitled to make an election. It will then be treated as having effect from the date on which the second residence was acquired (HMRC Extra-Statutory Concession D21). This concession is intended to cover the position where one of the properties is rented and the taxpayer did not realise that it constitutes a residence.

SEPARATION OR DIVORCE

A married couple can only have one main home between them. Where a married couple separate or are divorced and one partner, say the husband, moves out and subsequently, as part of a financial settlement, disposes of the home (or his interest in it) to his wife, it can be regarded as continuing as the husband's main residence during the period up to the date of the transfer, provided that it continues to be the wife's main residence throughout that period and the husband does not have any property which he is claiming to be his main residence (HMRC Extra-Statutory Concession D6). It should particularly be noted that this applies only where the house is ultimately transferred to the wife. If it is sold and all or part of the proceeds are paid to her, the concession will not operate. In such a case, it may be preferable to transfer the property to the wife for her to sell.

BUYING OVERSEAS

An expatriate who buys a UK house while overseas would not have met the rule of using the house both before and after a period of absence as the principal private residence. Instead, the expatriate should consider selling the house in the tax year before returning, while non-resident and therefore exempt from all capital gains tax (although there may be tax consequences in their country of residence). If you work full-time overseas, HMRC will regard you as non-resident until the day you return to the UK, in which case it would appear unnecessary to sell the house in the tax year before you return; selling before you return, even in the same tax year, would be sufficient. However, if possible you should not rely too heavily on this, in case your residence is deemed to start earlier.

Spanish capital gains tax

A disposal of a house (or any other property) in the UK is liable to Spanish capital gains if you are resident in Spain at the date of disposal. The current CGT rates in Spain are 21% (on the first €6,000 of combined gains and investment income), 25% on the next €18,000 and 21% thereafter.

18.5 Succession law and taxes: UK and Spain

Succession Law Applicable

The applicable succession law will be UK law.

UK inheritance tax

Property in the UK owned by an individual remains liable for UK inheritance tax on the death of the owners, regardless of domicile. A non-UK domiciled individual can avoid UK inheritance tax by placing the property into an offshore company. Further advice should be sought as there are UK tax consequences of such a structure.

Spanish succession tax

Property outside Spain remains liable for Spanish succession tax if gifted to or inherited by a resident of Spain. If gifted to or inherited by a non-resident of Spain, it will be outside the scope of Spanish succession tax.

CHAPTER 19

UK Inheritance Tax (UK IHT)

19.1 Introduction

UK inheritance tax (IHT) is payable on estates of UK domiciled individuals and on gifts made by UK domiciled individuals within the seven years prior to death. UK assets are also always liable to UK inheritance tax.

IHT is payable on gifts into most types of trust made by UK domiciled individuals.

If you are non-domiciled in the UK at the date of death or gift, UK inheritance tax is only payable on your UK assets.

An affidavit certifying the assets you own has to be submitted to HM Revenue and Customs (HMRC) by your personal representatives or the donees or trustees and the tax paid (with some exceptions) before a grant of probate of your Will can be obtained, and therefore before any of your assets can be distributed to beneficiaries.

Spanish succession tax is payable on gifts or inheritances received by Spanish residents, and on assets situated in Spain, **see Chapter 8**.

There are a number of problems and opportunities which can arise in certain circumstances.

19.2 Rates of UK IHT

IHT bites quickly and hard. It is not only a tax for the very rich, as quite modest estates are liable to the tax.

UK rates on death until 2017/18

Gross transfers	Rate %
£1 - £325,000	0
Over £325,000	40

Thus, the tax on a £500,000 estate is £70,000.

The amount taxable at nil per cent is generally referred to as the 'nil rate band'. There are technical reasons why the Government has chosen to impose tax at nil per cent rather than exempt the first £325,000 from tax.

The nil rate band of £325,000 per year has been frozen up until at least the 2017/18 tax year.

The rate of tax on a lifetime gift which is immediately chargeable to tax (eg, a gift into a trust) is 20% (half the death rate).

19.3 The surviving spouse or civil partner

IHT is not usually payable when assets pass to a surviving spouse or civil partner. Accordingly, if husband and wife leave all their assets to one another, no IHT will be payable until the second death. There is, however, a limit to this exemption where assets pass from a UK domiciled spouse to a non-UK domiciled spouse (see below).

It is possible for spouses and civil partners to transfer any unused nil rate band allowance to the surviving spouse or civil partner for use on the second death. The transfer is available regardless of when the first partner died/dies.

The claim to transfer any unused nil rate band must be made when the surviving spouse or civil partner dies and not when the first spouse or civil partner dies.

Where a valid claim to transfer unused nil rate band is made, the nil rate band that is available when the surviving spouse or civil partner dies will be increased by the proportion of the nil rate band unused on the first death. For example, if on the first death the chargeable estate is £112,500 and the nil rate band is £325,000, 50% of the nil rate band would be unused. If the nil rate band when the survivor dies is £325,000, that would be increased by 50% to £487,500.

The amount of the nil rate band that can be transferred does not depend on the value of the first spouse or civil partner's estate.

Whatever proportion of the nil rate band is unused on the first death is available for transfer to the survivor.

The unused nil rate band can be transferred from more than one deceased spouse or civil partner, up to a limit of one additional nil rate band. So if someone has survived more than one spouse or civil partner, then on their death a maximum of 100% of the nil rate band in force at that time can be transferred.

Up until 5 April 2013, if one spouse is domiciled in the UK and the other is not, the usual spousal exemption is limited where the non-UK domiciled spouse inherits from the UK domiciled spouse. Only the first £55,000 of assets passing to the surviving spouse is exempt in such a case (and even that is reduced if gifts were made to the spouse in the previous seven years). From 6 April 2013:

- The cap of **£55,000** is increased to the current nil rate band of **£325,000**; *and*
- Non-UK domiciled spouses will be given the option to **make an election to be treated as UK domiciled** for IHT purposes.

The election can be made during the lifetime of the UK domiciled spouse, or within two years of their death. This will enable non-UK domiciled spouses to receive assets from their UK domiciled spouses on death, free of tax. The disadvantage of making this election is that the worldwide estate of the non-UK domiciled spouse is taken into consideration for IHT purposes. If no election is made, then only the UK assets of the non-domiciled spouse would be subject to IHT.

Where the election is made after death, where the non-UK domiciled spouse is non-UK resident, or will become so shortly, the election will only last for four years, so after the expiry of the election, all non-UK situated assets will fall out of the UK inheritance tax net. Otherwise, the election remains in place for life.

19.4 Lifetime gifts

Where an individual dies within seven years after making a lifetime gift, this may give rise to a charge (or an increased charge) to tax. The exact nature of the charge depends upon whether or not the lifetime gift was a 'potentially exempt transfer' (PET), ie, a gift on or after 18 March 1986 to an individual, or a 'chargeable lifetime transfer' (CLT), ie, the creation on or after 22 March 2006 of a trust (discretionary or life interest).

You may give away unlimited sums to other individuals during your lifetime, subject to your survival for seven years after the date of the gifts (known as a Potentially Exempt Transfer or PET). In this instance, the value of the original gift and any growth in the value of the gift in the interim seven-year period will not form part of your estate for IHT purposes. The gift must be absolute, ie, there must be no possible reservation of benefit for yourself, or the asset will remain in your estate. If the lifetime gift was a PET, the effect of death within seven years is that the gift becomes chargeable to tax at the rate which would have been applicable had the gift been made on death.

At the time when a potentially exempt transfer is made, no tax is payable. It is assumed that the transfer will prove to be an exempt transfer. This assumption is maintained until it is either proved correct, by the survival of the donor for seven years, or incorrect, by the death of the donor within the seven-year period.

If the donor fails to survive the seven-year period, the gift becomes a chargeable transfer as at the date of the gift but is taxed using the rates and nil rate band as at the date of death – with 'taper relief' if appropriate. To calculate the tax chargeable on the gift, the following steps must be followed. (NB: these steps take account of changes in tax rates between the gift and the death.)

STEP 1: Ascertain the transferor's cumulative total immediately before the potentially exempt transfer was made. The cumulative total will comprise:

- Chargeable transfers made within seven years before the gift; *and*
- Potentially exempt transfers made within that period and within seven years before the death (ie, PETs made more than seven years before death are not included here).

STEP 2: Calculate the tax payable on the transfer, using the rates current at the date of death, taking account of the cumulative total at Step 1.

STEP 3: Apply taper relief if the gift was made more than three years before death (see below).

There is a taper relief as regards gifts made more than three but less than seven years before death, which reduces any tax on those gifts as follows:

	Tax due
Years 1 - 3	100% of normal tax due
Year 4	80% of normal tax due
Year 5	60% of normal tax due
Year 6	40% of normal tax due
Year 7	20% of normal tax due

The saving obtained with taper relief can be significant.

If the donor survives the seven-year period, the gift becomes an exempt transfer. No tax is payable on it, and it drops out of the deceased's cumulative total for the purpose of calculating tax on their estate when they ultimately die.

If the lifetime gift was chargeable to tax at the time the gift was made, eg, because it was a gift into a trust, the effect of death is that extra tax is payable because the gift is now chargeable at the full death rates (40%), instead of the half death rates (20%) at which it has been paid.

The first £325,000 passing from a UK domicile to a non-UK domicile is not treated as a PET, but will use up the lifetime non-domiciled spouse exemption. If so used, this exemption will not be available on death

19.5 Introduction to domicile

Domicile is a concept of UK common law. Not all countries have the concept but those who do generally have different rules to the UK. The major tax effect in respect of this concept is that those domiciled in (or deemed to be domiciled in) the UK are fully liable for UK inheritance tax upon their worldwide assets.

While an individual may be resident in more than one country, under UK law, an individual may only have one domicile at any single point in time. An individual's domicile is not necessarily the same as either their residence or nationality.

Residence is a transitory concept. It is easy to change one's residence and it is also possible to be resident simultaneously in more than one country.

Domicile is a different matter. The basic rule is that a person is domiciled in the country in which they have their home permanently or indefinitely. This is often described as the country which the person regards as their 'homeland'. This is frequently described as the place where a person intends to die, although it is the whole picture of an individual's life which is relevant, not just one single factor.

Your place of domicile does not have to be the country with which you have your closest personal association. A person can live in a country for many years and still remain domiciled elsewhere.

English law recognises three kinds of domicile:

1. Domicile of origin.
2. Domicile of dependence.
3. Domicile of choice.

Acquiring a new domicile is considered to be a very serious change by both HM Revenue and Customs and the UK courts. It is the party alleging the change of domicile that bears the burden of proof (being the civil law burden of the balance of probabilities), and the UK tax authorities will not usually make a domicile determination unless tax is at stake. So unless you settle more than the nil rate band – £325,000 until the end of the 2017/18 UK tax year – or die, you are unlikely to find out what they consider your domicile to be.

Given that it is usually only after your death that any claim to have acquired a domicile of choice overseas is made and challenged, in which case it will usually be up to your heirs to prove any change of domicile, it is prudent to take advice on the matter, and to document this as much as possible.

19.6 Domicile of origin

At birth, an individual acquires a domicile of origin.

A child born legitimately has a domicile of origin of the father's domicile at the time of the child's birth. Note that this is not necessarily the same as the country in which the child was born. Indeed, it is possible for someone to have a domicile of origin in a country which they have never visited.

A child born illegitimately, or born legitimately but after the father's death, takes the mother's domicile at the time of birth.

A foundling takes on the domicile of the country in which they are found.

In the event of a child becoming adopted, a new domicile of origin will be acquired since adoption involves the complete severance of the legal relationship between parent and child, and the establishment of a new one between child and adoptive parent.

Should an individual acquire a domicile of dependence or a domicile of choice, the link to domicile of origin would be removed, but not destroyed; rather, it will be held at readiness to be re-attached to them instantly should their domicile of dependence cease, or their domicile of dependence or choice be abandoned.

The domicile of origin never changes and it revives automatically if there is any doubt about an individual's domicile of choice or dependence. Note that an individual can never be without a domicile under UK law.

19.7 Domicile of dependence

Until 1 January 1974, there were three classes of persons who would acquire a domicile of dependence: children, mentally disordered persons, and married women. Only the first two classes remain; women who married after 1 January 1974 do not assume the domicile of their husband.

For example, a Spanish woman marrying an English man before January 1974 would automatically assume an English domicile. After that date, a woman already married will retain her husband's domicile until it is changed by acquisition or revival of another domicile. In this example, if the couple moved to live in Spain, she would quickly acquire her former Spanish domicile, and abandon her English domicile. Her husband might retain his UK domicile, however. For marriages after 31 December 1973, the wife's and husband's domicile is determined separately. There is also a general rule whereby any UK domicile who alters their domicile remains treated as UK domiciled for the purposes of UK IHT for three years after the change.

There is nothing to prevent a domicile of origin being displaced if the domicile of your own parents alters. Before 1 January 1974, this would happen if the parents altered their domicile before the child reached 21 years of age. Since 1 January 1974, it happens before the child reaches 16 (or marries, if earlier).

19.8 Domicile of choice

The domicile of origin or dependence can be changed to a domicile of choice. However, this is not easy as there is usually a strong presumption in favour of retention of domicile of origin. Furthermore, if you lose your domicile of choice, eg, by leaving your adopted country and moving to a third country, you are likely to reacquire your domicile of origin, unless you can demonstrate that you have moved to a new domicile of choice.

Until 1 January 1974, it was the rule that children under the age of 21, or not married, could not acquire a domicile of choice. Since then, any child reaching the age of 16 or marrying under that age, has been capable of acquiring a domicile of choice.

There are two main elements for a domicile of choice to replace a domicile of origin. First, the individual must be physically present in the new country and be a tax resident there. Second, an intention must be formed to reside permanently in the new jurisdiction.

The sort of things that HMRC look at to consider where the person is domiciled are: the country of which they have a passport, or a driving licence, or in which they have made a Will, arranged to be buried, are a club member, hold credit cards, personal insurances, bank accounts, property and have family.

If you are domiciled in the UK then UK IHT will be payable on your worldwide assets at the date of your death, within normal UK IHT rules. The first £325,000 is tax-free and the remainder is taxed at 40%. Transfers between spouses who are both UK domiciled, or who are both non-UK domiciled, are exempt.

For most retired expatriates it is normal to assume residence of the country in which you have retired while continuing to retain UK domicile status. This, in itself, poses a number of complex problems.

Most important, in the event of your death difficulties can arise in respect of the jurisdiction in which death duties are to be paid.

19.9 Abandonment of domicile

A person abandons a domicile of choice in a country by ceasing to reside there and by ceasing to intend to reside there permanently or indefinitely and not otherwise. It is necessary for both the residence and intention, which must exist for the acquisition of a domicile of choice, to be relinquished. Thus, if you leave Spain, for example, and do not intend to reside there permanently or indefinitely, you will be seen to have abandoned your Spanish domicile of choice.

When a domicile of choice is abandoned, two possibilities arise:

1. A new domicile of choice is acquired; *or*
2. The domicile of origin revives.

19.10 Acquisition of a new domicile of choice

On abandoning a domicile of choice, a person may acquire a new domicile of choice. This is established in accordance with the rules described above. Thus, a domicile of choice is acquired by the combination of residence and intention of permanent or indefinite residence. For this purpose, residence means physical presence as an inhabitant of a country. However, as a matter of law, it is not necessary that residence should be long in point of time. Residence for a few days or even a day is sufficient. An immigrant can acquire a domicile immediately upon arrival in the country in which they intend to settle. Length of residence is not important in itself. It is only important as evidence of intention to remain. The standard of proof required to establish a change of domicile of choice is on the balance of probabilities, and the burden falls on the party who is claiming the change of domicile.

It should be noted that under English law, an individual cannot be domiciled in a federal state, but only in one of the states.

19.11 Revival of domicile of origin

On the basis that you presently have a domicile of choice in Spain, for example, your domicile of origin will only revive if you leave Spain and no longer intend to live there permanently, but you do not have the intention of living in the new country permanently or indefinitely, or if you return to your domicile of origin.

19.12 HMRC's approach

So, what if a living UK resident wants to prove that they are non-UK domiciled? Paradoxically, HMRC will not agree to an individual's domicile unless a current tax liability is in issue.

As non-UK domiciled individuals are only liable for income tax and CGT on the remittance basis, anyone in doubt as to their domicile status could declare foreign income or gains to HMRC as part of their self-assessment return.

HMRC might then make an express statement as to whether or not the individual is UK domiciled. Should a formal tax enquiry be necessary, form DOM1 will be issued by HMRC for the taxpayer to complete and will decide the matter for that tax year.

19.13 The '17 of the past 20 years rule'

Even if an individual is non-UK domiciled and making use of the remittance basis for UK income and UK capital gains tax (and paying the corresponding £30,000/£50,000 charge), the deemed domicile rules for inheritance tax purposes will come into effect after UK residence for 17 of 20 fiscal years.

19.14 The UK domicile maze: UK law

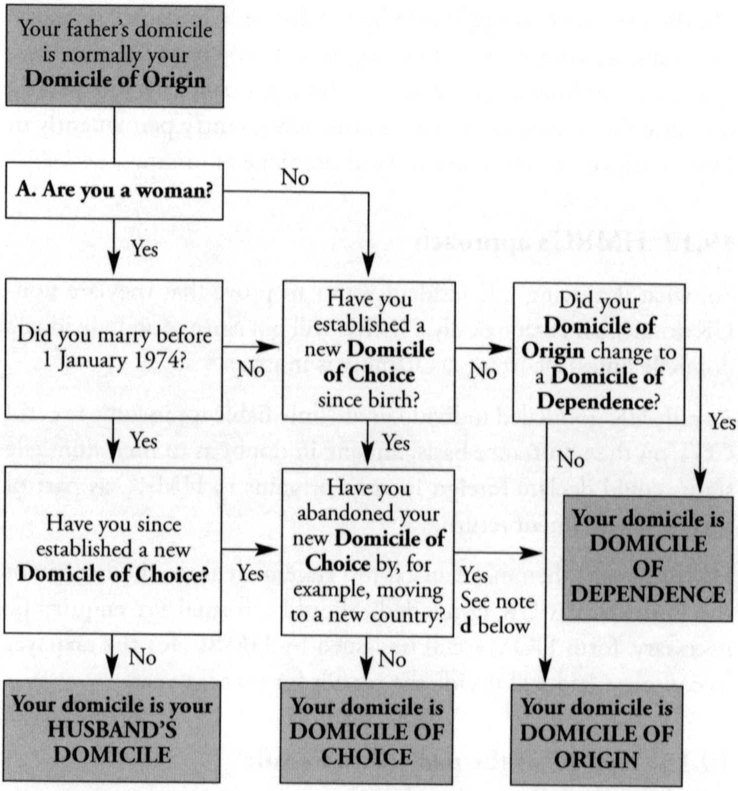

NOTES

a. If you have been resident in the UK for any part of 17 of the previous 20 UK tax years, you are deemed UK domiciled for UK IHT purposes regardless of the above. In determining 'residence' for this purpose, HMRC has confirmed that the old available accommodation rules can be ignored.

b. If you have changed your UK domicile you are deemed to remain UK domiciled for UK IHT purposes for three years after the change.

c. There is no UK inheritance tax exemption for transfers from a UK domiciled spouse to a non-UK domiciled spouse, other than an exemption limited to £325,000 which, when added to the current nil rate band of £325,000 (for 2013/14) totals £650,000.

d. If you have moved to a new country with the intention to reside there permanently, you would establish a new domicile of choice. If not, your domicile would revert to your domicile of origin.

"If you master your destiny, you will find out there is a destiny tax."

JOSEPH BONKOWSKI

UK Inheritance Tax

The UK has a long reach, sometimes crossing international boundaries! When it comes to UK inheritance tax, it is worth exploring the implications for your specific personal situation.

We need to talk.
www.blevinsfranks.com

CHAPTER 20

UK Companies Paying Dividends to Spanish Residents

20.1 Introduction

UK dividends have a UK tax credit attached to them of 10%. Therefore, the actual amount you are treated as having received is the dividend received (the net amount) plus the tax credit (the gross amount). So, if you receive a dividend of £90, the tax is £10, so for tax purposes, you are treated as having received £100 in total.

The tax credit is not repayable in the UK, as this is effectively notional tax; that said, it can be used to offset your tax liability on this income both in the UK and Spain.

20.2 Effect on UK residents

Lower or basic rate individual taxpayer

There will be no further tax to pay on the dividend. The dividend is treated as if there was a 10% tax credit on the gross dividend (ie, one-ninth of the net dividend received), and this will satisfy in full the UK income tax liability of a lower or a basic rate taxpayer. Thus the effective rate of income tax for this group is 10%.

Higher rate taxpayers

Individuals who pay tax at 40% in the UK will have an additional income tax liability but the rate of tax on dividend income is reduced to 32.5% against which the 10% credit is set. This equates to further income tax of 25% of the net dividend received.

Individuals who pay tax at 45% in the UK pay tax at a rate of 37.5% on dividend income, with a 10% tax credit. This equates to further income tax of 30.6% of the net dividend received.

Non-Taxpayers

The tax credit will not be repayable to non-taxpayers and no tax credit is available where the income falls within the UK personal allowance.

20.3 Effect on non-UK residents

UK dividend income may be treated for UK tax purposes in one of two ways in the hands of non-UK residents, depending on whether UK personal allowances are claimed, where these are available.

British citizens, European Economic Area (EEA) citizens, Crown Servants, and certain others, may all claim UK personal allowances against UK-taxable income, regardless of their residence position in the UK.

Even as a non-resident you can still potentially be a higher-rate taxpayer in the UK if your UK income is high enough.

The rules say that the total UK income tax of a non-resident cannot exceed:

- The tax that would be charged if 'disregarded income' (eg, dividends, see below) and personal allowances were ignored; *plus*
- The tax deducted or treated as deducted at source from the 'disregarded income'.

What this means in practice is that you can entirely avoid UK higher rate tax on substantial UK dividends as a non-resident, but other taxable income such as UK rental income might then be taxed without the benefit of personal allowances. It depends on the figures involved as to which works out better. But either way the eventual UK tax to be paid cannot exceed that on non-disregarded income without any allowances.

HM Revenue and Customs (HMRC) usually does the comparative calculations automatically and gives you the most beneficial treatment, but it is worth checking that this has been done.

The effect of the two treatments is described below.

Claiming the UK personal allowance

If it is better to have your UK personal allowances (eg, because you have UK-taxable income such as rental or government service pension) and you are a UK lower or basic rate taxpayer, the 10% tax credit treated as being paid at source will satisfy the UK tax liability on the UK dividend income, and therefore there will be no further UK tax liability due. UK higher rate taxpayers are liable for paying a further 22.5% of the gross dividend received (25% of the net), and additional rate taxpayers 37.5% (30.6% of the net).

When calculating UK income tax on the income, no credit is given in the UK for dividend income falling within the UK personal allowance.

Not claiming the UK personal allowance

Where you have no need to claim your UK personal allowances (ie, because your income only consists of the 'disregarded income' items listed below) UK dividend income (and all other disregarded income) will only be subject to UK tax to the extent that income tax is deducted at source, if at all. If this is the case, the 10% dividend tax credit will satisfy the UK income tax liability on the dividend income, regardless of your lower, basic or higher rate tax status.

Therefore, claiming personal allowances is only likely to be beneficial where you have other UK taxable income (eg, UK sourced rental income or government service pension taxable in the UK) which is relieved wholly or partially by the personal allowance, and where other income that would otherwise be 'disregarded', such as UK dividends, is not substantial.

'Disregarded income' includes:

- Dividends from UK companies (these can never be paid gross as the 10% tax credit is not real but a deemed amount and the UK tax liability will always therefore be maintained at the credited 10% level).
- Income from unit trusts (again with a 10% tax credit if equity-based).
- Interest.
- Certain social security benefits such as state retirement pension, widowed parent pension, or other taxable state benefits (usually paid gross, ie, no tax deducted at source).
- Certain retirement annuities, but not most personal or occupational pensions.
- Taxable income from purchased life annuities (except those under personal pension or trust schemes).
- Profits from public revenue dividends.
- Profits or gains from transactions in deposits.

20.4 Spanish tax treatment of UK dividends

Dividends are subject to tax at rates of 21%, 25% and 27%. The 21% rate applies on the first €6,000 of combined investment income and capital gains. Where such income and gains fall between €6,000 and €24,000, tax is payable at 25%, for income and gains over €24,000 the tax rate is 27%. An allowance of €1,500 per annum is usually available.

A resident of Spain who receives dividends from a company resident in the UK is entitled to the same tax credit that an individual resident in the UK would be: ie, 10%. This tax credit can be offset against the Spanish tax liability on the dividend income.

The taxable dividend in Spain would be the net dividend received **plus** the 10% tax credit treated as being attached to the dividend (see example below). Thus, in Spain, a further 11%, 15% or 17% will be due in respect of UK dividend income.

An example is set out below for a net dividend of €10,000 received from a UK company:

	€
Gross dividend (€10,000 x 100/90)	11,111
Less: allowance	(1,500)
	9,611
Spanish tax thereon*:	
€6,000 x 21% = €1,260	
€3,611 x 25% = €903	2,163
Less: UK 10% tax credit	(1,111)
Tax payable in Spain	1,052
Effective tax rate	9.47%

**Assumes no other investment income or gains received during the year.*

If the shares have been placed into an offshore company/trust structure, or the dividend received before becoming a Spanish tax resident (and in the UK tax year after leaving the UK), there are no further UK or Spanish taxes on the dividend.

CHAPTER 21

Wills and Spanish Succession Law

21.1 Why have a Will?

It is important to make a Will or Wills if you have assets in different countries. On your death a Will ensures that your estate will pass to your beneficiaries of choice and reduce potential time delays in distributing your estate. Careful estate planning through the use of a Will may enable many people to reduce inheritance taxes in both the UK and Spain.

What happens to your estate when you die depends on your Will (if any) and the laws of the two countries. Although the rules are complex, in practice if you make adequate provision for your immediate family in a Will drafted under either country's laws, there should not be a problem. It is normally preferable to have two Wills: a Spanish Will for any assets in Spain and the relevant UK Will for everything else. However, this largely depends on whether your beneficiaries are resident in Spain or not.

21.2 What happens if I do not make a relevant UK Will?

The consequence of dying without a Will can be serious. A relevant UK Will made before a marriage, or re-marriage, is normally rendered ineffective by the change of circumstance: you have to make another one. If you die without a Will and are UK domiciled (remember that many UK expatriates who are resident in Spain and have UK assets will remain UK domiciled) there are set rules as to where your estate goes (called the Intestate Succession Provisions). The advantage of having a Will is that your estate will be distributed as you want it to be, the administrators of the estate will be chosen by you, and the estate is likely to be distributed faster and at lower costs. The statutory rules can give rise to an unnecessary inheritance tax liability. For example, many people believe that if they die without a Will, all of their assets will pass to their widow/er. This is not the case unless the total assets are less than £125,000 if you have children, or less than £200,000 if you have no children.

If you die without having made a Will, and are domiciled in UK, Wales or Northern Ireland, at death your assets will be distributed according to the rules of intestacy. These rarely reflect what you are likely to want to happen to your estate. The intestacy rules in UK, Wales and Northern Ireland are set out below. Note that the terms widow/widower also include civil partners.

Deceased is married with no children

The widow or widower receives:

- All personal possessions.
- First £200,000.
- One half of everything else.

The remaining half goes to the deceased's parents. If no parents are alive, the individual's brothers and sisters (or their issue) will share the remaining half-share. If there are no such relatives, the widow or widower will take the entire estate.

Deceased is married with children

The widow or widower receives:

- All personal possessions.
- First £125,000.
- A life interest in one half of everything else.

The remaining half passes to the deceased's children in equal shares. They will also inherit the first half when the surviving spouse dies. The surviving spouse may redeem the life interest, ie, surrender it in return for a part of the capital when all the beneficiaries are adults, so the whole estate may then be distributed.

If the deceased leaves children but no spouse, the estate is divided among the children equally.

Unmarried, no children

The entire estate passes to the parents. If no parents are alive, the estate passes to the nearest surviving relatives in the following order:

- Brothers and sisters.
- Children of above.
- Grandparents.
- Aunts and uncles.
- The Crown.

If anyone is alive in the first category, the whole estate is divided between those people. If there are none, it is instead divided between everyone in the next category who is alive, and so on.

Common law partners

They do not benefit under the intestacy rules in the UK, though anyone dependent on the deceased can make a claim for financial provision under the Inheritance (Provision for Family and Dependants) Act 1975.

21.3 Dying without a Will: Scotland

Where an individual domiciled in Scotland dies without a Will, the executors (having paid the deceased's debts) have to distribute the estate in the following order.

Firstly, prior rights exist in favour of the surviving spouse. There are three such prior rights:

1. The deceased's dwelling house, if the surviving spouse ordinarily lived there. In some cases the spouse receives a cash sum instead, eg, if the house exceeds £65,000 in value.
2. Furniture and belongings of the deceased, if situated in a house where the surviving spouse lives. If these are worth more than £12,000 the spouse receives cash instead. Heirlooms are excluded.
3. A cash sum from the estate remaining after 1 and 2 above have been satisfied. This is £21,000 if the deceased left surviving children; £35,000 if there are no children.

If any estate remains, the claim with next priority is the legal rights in favour of the surviving spouse and children. These are automatically protected and cannot be overturned by a Will. These legal rights are described above.

Having satisfied the legal rights, the remainder (if any) is distributed to the children, or if there are no children, to the parents and brothers or sisters, or just to the brothers or sisters (if no parents are alive), or to the parents only (if no brothers or sisters are alive). If there are no children, parents, or brothers or sisters, the surviving spouse will receive everything.

21.4 Dying intestate in Spain

Dying without a Will can cause time-consuming and expensive legal procedures for your survivors, and may not achieve how you want your estate distributed as it will be distributed according to the Spanish laws of succession.

21.5 Spanish Wills

As in the UK, it is not possible to establish joint Wills in Spain, so each individual in a couple will each have to establish separate Wills.

The process for writing a Spanish Will is different from that of a relevant UK Will. Valid Spanish Wills are usually prepared by a Spanish lawyer, or other legal representative, and then must be signed in front of a notary. For UK expatriates, the Will is prepared in both Spanish and English, although it is only the Spanish part that is regarded as the legal Will. To this extent the solicitor will be present at the signing of the Will at the notary's office to ensure that you have fully understood the Will before signing. In some instances a legal or 'fiscal agent' will carry out this function.

Once the Will has been signed, the original will be kept in a central Will registry in Madrid and you will be given a copy for safe-keeping.

European Certificate of Succession

From 7 August 2015, under EU Regulation 650/2012, the general rule across the EU countries that have signed up (all EU states excluding the UK, Denmark and Ireland) will be that the law of your country of habitual residence will apply to all of your assets, including real estate held within a different EU country.

Habitual residence is not defined in the regulation, but the country dealing with the succession will look at all of your circumstances at death and in the preceding years to establish the country of habitual residence. This will include countries outside the EU.

However, an individual may instead opt for the laws of his nationality to apply instead, even if this is a country outside the EU. In order to opt for the laws of a country of which he is a national, the choice must be made before death.

These laws do not change the tax position on death – if you are resident in Spain at the date of death, the Spanish succession tax rules will continue to apply.

21.6 General points on Wills

When drafting Wills in more than one country it is important that this is done carefully, so that each Will does not compromise another. For example, an individual establishes a UK Will and subsequently establishes a Spanish Will. If each Will does not recognise the existence of the other, then it is possible that the UK Will becomes invalid and from a UK perspective the deceased may be deemed to have died intestate.

Professional advice should be taken to ensure that all of the estate is properly distributed and partial intestacy is avoided. Careful consideration should be given to what would happen if the person to whom you are leaving your estate were to die just before or just after you. You should also be aware that in the UK a dependant has a right to make a claim against the estate if reasonable financial provision is not made in your Will.

21.7 Trusts

You may wish to create a trust on your death if, for example, you want the income to go to your spouse but the capital to go to your children when your spouse dies.

A special kind of UK Will, called a discretionary Will trust, enables you to alter your Will without going through all of the formalities of drawing up a new one as circumstances change. The usefulness of this technique in your particular circumstances should be carefully discussed with your professional advisers.

If you have put assets into trusts during your lifetime, these assets are not governed by your Will; they will continue to be dealt with in accordance with the trust deed. However, when deciding how you want your estate to be divided on your death, do not overlook such trusts; you may want to co-ordinate the two. For example, if you want to leave substantial legacies to people other than your spouse, it may save IHT if you make them beneficiaries of your settlement and provide in your letter of wishes to the trustees that payments should be made to them on your death. You may want to include them in your Will in case there are insufficient assets in the settlement at that time. If so, you will want to provide that they should benefit under your Will only if, and to the extent that, they do not benefit under the settlement.

If you have assets situated in other countries, you also need to take account of the inheritance laws of those countries, and the tax treatment in those countries of the assets within the trust. Specific advice should be sought.

21.8 Spanish succession law

Most continental European countries, including Spain, apply the Napoleonic Civil Code, whereby the children of the testator have an automatic right to inherit a part of the estate on death. This applies even if they are not expressly mentioned in your last Will and testament.

Spanish succession laws restrict your freedom to leave your assets to anyone you please. This is to protect the family and provide for the children. The Spanish laws require a parent to leave two-thirds of their estate to the children, even by-passing the surviving spouse.

You may find that your plans agree basically with what Spanish law provides if you wish to provide for your children before the second death, and you might have no problem in making a Spanish Will in accordance with those provisions.

In Spanish law, a surviving spouse keeps:

- All assets acquired before the marriage;
- Half of the goods acquired during marriage; *and*
- All personal gifts or inheritances which have come directly to them.

If most of the couple's assets were acquired during their marriage, then normally half of the assets (or thereabouts) continue to belong to the surviving spouse. This is particularly true of jointly owned real estate. The surviving spouse continues to own their half share, with their name on the title deed.

Of the rest of the assets, only one-third can be freely disposed of (ie, one-sixth of the combined total if it was all owned jointly).

When a person dies leaving children, the estate is divided into three equal parts.

One-third must be left to surviving children (or more remote issue) equally.

Another third must also be left to the children, but the testator can decide how to divide this up. For example, he could choose to leave all of this third to only one child or grandchild. A surviving spouse has a life interest (usufruct) over this third.

The final third of the estate can be freely Willed to anyone.

If the children do not wish to claim their Spanish legal share of the inheritance, they can renounce their entitlement in front of a Spanish notary, and the estate will then be distributed amongst the beneficiaries designated in your last Will.

21.9 Application of Spanish succession law to non-Spanish nationals

The Spanish succession laws in theory apply to foreigners with property in Spain.

However, Article 9 of the Spanish Civil Code states that when a foreigner owning Spanish property dies, (even if they hold an official residence permit) the disposal of any assets in Spain will be governed by the law of the country of nationality, not Spanish law. If your own country's law permits free disposal of the estate, this frees you from the Spanish law of reserved heirs. It does not free the inheritors from Spanish inheritance tax (succession tax), though.

This freedom applies only when the foreigner has an existing foreign Will or Spanish Will. If they die intestate, without having made a Will, Spanish law will be applied to their assets in Spain and these will be divided equally among their children.

21.10 Application of Spanish succession law to UK nationals

For UK nationals, a complication arises.

British law states that the disposition of real estate is to be governed by the law of the country where the property is located.

So, Spanish law says that UK law (the law of nationality) will apply, but UK law says that Spanish law will apply to a property located in Spain. This is called a double *'renvoi'* (reversion to another country's law) and so, a UK national in theory is subject to the Spanish law –

which **might** mean they can freely dispose of only one-third of their assets in Spain even if they have a Will.

However, in practice, a UK national can make a Spanish Will bequeathing their Spanish property to whomever they like as their own national law allows free disposition of property, provided they make a Spanish Will. The Spanish Registrar of Wills has in the past accepted this, and when the time comes, the Will is executed accordingly. Spanish lawyers routinely make such Wills.

This is largely because despite the double *renvoi* mentioned above, the Spanish courts will for practical purposes prefer to apply unity of succession law, and not two different systems (say, the Spanish and the British) to the one deceased's estate.

So most foreigners will find no problem in making a separate Spanish Will to dispose of their property in Spain, whether they are residents or non-residents.

It is therefore important to have both relevant and valid UK and Spanish Wills if you are either Spanish resident or UK resident with Spanish assets.

21.11 Warning – legitimate inheritors could contest a Spanish Will

All of the above works perfectly well in most cases. But, it is not exactly in agreement with the law. It is more a matter of the Spanish authorities choosing not to enforce their own laws too strictly. This means that, if you write your Spanish Will, cutting out one or more of your children, they could in principle contest the Will under Spanish law on the grounds that the law stipulates that part of the inheritance is theirs, and win.

If you foresee any possible successful challenge from a legitimate inheritor under Spanish law, such as a child or a spouse, you should make other arrangements during your life, such as transferring the title of the property to your chosen heir while you are still alive. This will trigger gift tax at that point, but this may be a small price to pay if it ultimately achieves what you want. You can always reserve the usufruct over the property, which gives you the right to use it for the rest of your life, although the underlying title *(the nuda propiedad)* has passed to another person.

21.12 Spanish supreme court decision 2004

The spotlight was thrown on to this problem by a 2004 Supreme Court decision in Spain that seems, at least, to say that if there are no UK assets the Will may be subject to Spanish law. In this instance, children successfully laid claim to their reserved inheritance of a Spanish property that had been left by a deceased UK national solely to his surviving spouse (their step-mother) under the terms of his Spanish Will. There were no UK assets nor, in this case, any other appreciable Spanish assets in the estate.

The complexity comes from the double *renvoi* created by the respective Spanish and UK private international law rules referred to above, and it makes sense in the context of Spain preferring to apply a unity of law. In this case there was no estate that ultimately fell under British succession law, there only being Spanish property which under the double *renvoi* is subject to Spanish law not British – and to follow that line did not offend the principle of unity.

21.13 Spain: procedure on death

On death, the surviving heir(s) must collect together a number of papers to deal with the inheritance. The '*Escritura* for the Acceptance of the Inheritance' must be signed and submitted to the property register and taxes paid, all within six months of the date of death.

The papers required are:

Original death certificate – this should preferably be an 'International Death Certificate' issued by larger registries and prepared in two or three different languages. If an International Death Certificate is not available, then you must obtain an *'Apostille'* (an internationally recognised legal certification) on the back of the ordinary death certificate and have it translated into Spanish by a sworn translator.

Original last Will and testament – once again, this Will ideally be their last Will and testament prepared in Spain. If they do not have a Spanish Will, then the original of their British Will with an *'Apostille'* will be required, along with a translation into Spanish by a sworn translator. They will also require a Certificate of Probate issued by the High Court in their country of nationality, with an *'Apostille'* and a Spanish translation by a sworn translator.

'Certificado de Ultimas voluntades' – issued by Madrid, confirming whether or not the deceased had properly registered a Will in Spain. The notary will normally apply for this certificate on receipt of the original death certificate. This generally takes approximately six weeks to arrive.

Escritura – or other proof of ownership of property left by the deceased. Remember for tax purposes that, as well as the house, the deceased person may also have owned other assets, such as plots of land, motor vehicle(s), bank accounts, etc.

As a beneficiary, all these items must be included in your inheritance declaration, otherwise you may have problems at a later date in selling the assets as they will not be in your name.

Once all the documents are assembled, the heirs must then sign the official *escritura* accepting (or renouncing) the inheritance in front of the Spanish notary. The value to be declared should be the approximate market value of the property or asset – too low a value may incur later penalties from the tax office.

Based on the value declared in the *escritura,* a self-declaration of the succession tax must be prepared by each beneficiary and the amount paid within the six months allocated after the date of death.

Under Spanish Law, should you 'forget' or be unable to claim your inheritance until after five years from the date of death, then the succession tax is no longer due. You may then claim your inheritance without paying any succession tax.

But, you would not be able to sell the property during this five-year period as the property is still in the name of the deceased person (even if owned jointly), and, if you are caught within the five years, tax, penalties and interest would be payable.

“Where there is a will, there is a way.”

SAMUEL SMILES

Wills

In order to ensure that your wishes for your family do happen, you need to have the right Wills in place in the UK and Spain.

We need to talk.
www.blevinsfranks.com

CHAPTER 22

Pensions and QROPS
(Qualifying Recognised Overseas Pension Schemes)

22.1 Introduction

There are many different types of pensions in the UK. The scheme you are a member of will depend, in part, as to whether you are employed or self-employed. Your entitlement to the state retirement pension will depend on the amount of national insurance contributions you made during your working life.

22.2 Company schemes

If you worked for a company and joined their pension scheme where contributions were made by you and your employer, it is likely that it was one of the following two schemes.

Money purchase scheme – A pension scheme where benefits are based on the fund provided by the contributions paid by you and your employer and the investment growth made on the contributions. This fund can never be in deficit.

Final salary scheme – A pension scheme where the benefits at retirement are based on the number of years that you worked for the company, and your salary at the date you left. The cost of providing the benefits should have been met by you and your employer, but such a fund can be in deficit in relation to its obligations, or in surplus.

22.3 Personal pension schemes

If you were self-employed, or worked for a company where there was no pension scheme, you would still have been able to save for your retirement by making contributions to a personal pension. Indeed, in some circumstances your employer may have also made contributions.

22.4 Some miscellaneous terms

Contracted Out – To be opted out of the State second pension (S2P) and State earnings related pension scheme (SERPS) and instead participate in a private, or employer sponsored pension arrangement which must provide certain benefits (see **Protected rights** below).

Protected rights – Benefits built up in a contracted-out personal pension or money purchase occupational scheme from the payment of minimum contributions and age related rebates by the Department of Work and Pensions (DWP).

Retirement annuity contracts (RACs) – A pension plan available to individuals who were self-employed or in non-pensionable employment. No new contracts were issued from July 1988 when personal pensions were introduced. However, increments can usually be made to existing contracts.

Lifetime allowance

Contributions made beyond the lifetime allowance do not attract tax benefits. The lifetime allowance is currently set at £1.5m but this went down to £1.25m from 6 April 2014.

The lifetime allowance applies even if you're in a pension scheme that you do not make payments into – for example if you're in a workplace pension that only your employer makes payments into.

Primary and enhanced protection: the new rules from 6 April 2006

Where you are drawing benefits now from pensions born before 6 April 2006, primary protection is available where the value of all pension rights accrued before 6 April 2006 exceeds the lifetime allowance of £1.5m.

It enables you to avoid (or at least minimise) the excess fund charge on funds in excess of £1.5m. The excess over the £1.5m at 5 April 2006 is expressed as a multiple of the lifetime allowance. The excess will be indexed in parallel with indexation of the lifetime allowance. When benefits are vested, only the excess over the multiple of the lifetime allowance will be subject to a tax charge.

If there was an excess fund, then the charge would be 55% but this is reduced to 25% if you transfer your pension under the QROPS rule (**see Section 22.9**).

Enhanced protection is also available to protect the capital value of all of a member's pension rights accrued before 6 April 2006, whatever their total value at that date or on vesting.

While enhanced protection ensures that all pension rights accrued before 6 April 2006 will be exempt from any tax recovery charge, this option is subject to members ceasing all pension contributions and active membership of all schemes from that date. Enhanced protection will protect the full value of tax-free cash lump sums, particularly if they are currently more than 25% of the value of the fund.

22.5 The annuity

The time has come for you to retire and take your benefits, so what options are available to you?

An annuity is a contract from an insurance company or mutual life office, which guarantees to pay an agreed level of income until the death of the insured life or lives. You pay a lump sum to an insurance company in return for a guaranteed income.

The annuity option may be applicable if you prefer to purchase security of income rather than take any further investment risk.

The main disadvantage is that on your death (or if you have an annuity for spouse or dependents, including a dependent partner, on the second death) the annuity ends and there is nothing remaining from it to pass to your heirs. There can be an exception if the annuity you buy is one which is guaranteed to be paid, regardless of whether you are alive or dead, for a period of time specified at the time of purchase. This is usually for five or 10 years.

If you take an annuity, you have some choices:

- Whether to take the cash commutation at outset and the amount you take (within certain limits). Cash commutation is a tax-free cash payment (in the UK it is tax-free) to you which reduces the pension fund and so reduces the amount of the annuity you can purchase.
- Whether the annuity is to remain level or will increase during payment.
- If it were to increase, whether the increase is tied to the retail price index or set at a fixed rate (eg, 3% per annum).
- Whether the annuity is to be paid for at least a minimum guaranteed term of years even in the event of an early death.
- If you buy an annuity which will have a minimum guaranteed period even if you were to die, the length of that period (five or 10 years, for example).
- Whether the annuity ends on your death ('single life') or is to continue to pay your spouse or dependent partner ('joint lives').
- If you were to buy a joint life pension, what level to apply (eg, the annuity for your spouse could be 50%, 75% or 100% of your own annuity).
- The frequency of your payment (for example monthly in advance, monthly in arrears, quarterly in advance, quarterly in arrears, annually in advance or annually in arrears).

These choices will affect the amount of your annuity when it commences.

Some options will have a significant effect on the size of the annuity. A good example is the difference between an annuity set up on a level basis, as compared to one that escalates while in payment, whereas the frequency of payment (monthly in arrears or in advance, etc) would have very little effect on the size of the annuity.

22.6 UK tax-free cash sum

Pensions can be partially exchanged for UK tax-free cash, subject to HM Revenue and Customs (HMRC) limits. The maximum depends upon the type of arrangement from which the pension evolves.

If a pension commencement lump sum is taken while you are resident in Spain, although it would be tax-free in the UK, it will be taxable in Spain.

22.7 Income drawdown before buying the annuity

Under current legislation, you can start to draw an income from your pension fund without having to buy the annuity; this is known as 'income drawdown'. It provides an ability to draw an income from investments held within the pension fund allowing personal investment control of the funds, as well as flexibility of income drawn.

This option may be applicable if you:

- Require income flexibility after retirement.
- Want to delay committing to an annuity (once bought, it is irreversible).
- Would like to take advantage of rising annuity rates (but you would lose out if they instead fell).
- Wish to maintain control of the investment strategy of your pension fund.
- Would like the opportunity to pass the pension fund to your heirs when you die.
- Need to hedge your pension against exchange rate fluctuations.
- Are prepared to accept some degree of risk with regard to your pension fund, in that the fund and future income stream are subject to investment fluctuations.

The specific decisions required at the outset are:

- Whether to take cash commutation at the outset and the amount.
- The initial level of income to be drawn from the pension fund.

It is usually possible to transfer existing pension arrangements to a new personal pension plan and draw your pension income from the pension fund itself, which remains invested in a portfolio of your choice, in a tax-efficient environment. The level of income can be varied each year and the actual purchase of an insured annuity can be deferred to a time of your choice in order to suit your personal circumstances.

You take the risk, however, that your investments may fall in value and/or that annuity rates (interest rates) decline.

Your options under income drawdown

Before 6 April 2011, you had to buy an annuity before reaching the age of 75 because of the tax treatment of the remaining fund on death (referred to as the 'left over fund').

The Government changed this position in Finance Act 2011. Since 6 April 2011, drawdown may continue beyond age 75. Benefits drawn down may be provided either directly from your pension fund through income withdrawal or in the form of a short-term temporary annuity (ie, up to five years duration).

The rules associated with drawdown:

GAD rates

The rates published by the Government Actuary's Department – known as the GAD rates – determine the extent of the amount that you may draw down by reference to notional annuity rates. The GAD rates may be found on the HMRC website.

The GAD rates vary by age and sex, and the correct GAD rate for a particular individual at a particular point in time is determined by their age, and the published yield from a particular investment.

The best way of going about deciding how much to draw down each year is achieved by following this process:

- Set the maximum permitted income that may be drawn down at 100% of the GAD rate applying.

  Require the maximum amount that may be withdrawn to be determined at least every three years until the end of the year in which the member reaches the age of 75, after which reviews will be carried out annually.

  If you have a secure lifetime pension income of at least £20,000 a year you can utilise flexible drawdown to gain access to some or all of your remaining pension fund without any cap on the withdrawals you may take.

- Set the tax rate on lump sum death benefits where benefit has been taken at 55%. But ensure that you nominate that in a situation where you have no living dependants the amount may be donated tax-free to a charity.

- Set the tax rate on lump sum death benefits payable beyond the age of 75 where the member has not taken benefit once again at 55%.

There are two forms of drawdown, flexible drawdown and capped drawdown. If you have secured pension income of at least £20,000 a year, you can access the remaining part of your pension fund by means of flexible drawdown. The amounts drawn down are subject to tax in the UK, but under the UK/Spain double tax treaty, the income is taxable solely in Spain, so you should be able to reclaim the UK tax paid on such income, although this process does take some time (usually between 18 months and two years, depending on the time of year the drawdown is taken).

If you have not secured pension income of at least £20,000 a year, and so are unable to gain access to the remaining fund, you may wish to utilise capped drawdown.

As suggested by the label, if you are in receipt of benefits where the capped drawdown rules apply, this is subject to maximum level of income that may be drawn down each year. That maximum is worked out by reference to the GAD rates. That maximum has to be reviewed at least every three years.

Cash commutation

You may still take part of your pension fund as a cash commutation, but it must be taken in its entirety when income drawdown commences. The maximum cash commutation is 25% of the fund value.

HOW UK PENSIONS ARE TAXED IN SPAIN

The taxation of pensions depends on what type of pension is being received. For tax purposes abroad, there are four different types of pensions:

1. State retirement pensions.
2. Occupational pensions.
3. Government service pensions.
4. Personal pensions.

UK pensions are always payable in Sterling. Some providers will make payments to a foreign bank, but many will not, so you need to check this with your provider. While you can have your pension transferred into Euros or any other currency, the exchange rate received will be that applying when the transfer is made. The foreign currency sum received can therefore vary, unless you arrange an agreed rate via a regular payments plan through a commercial foreign exchange dealer.

General points

The UK state retirement pension can be paid into either a UK or Spanish bank account. For other pension income, you should check the scheme rules, as some schemes prohibit payment into a non-UK bank account.

The UK generally taxes UK source pension income through the Pay As You Earn (PAYE) system, except for the state retirement pension which is always paid gross (but see below). However, provided that HMRC are satisfied that you are tax-resident in and paying tax in Spain, it is possible to arrange for the pension income to be paid gross (without deduction of tax at source).

It used to be possible to receive your UK pension without deduction of tax in the UK using Form FD9. However, this form is no longer used by the Spanish authorities and they will refuse to stamp any FD9s submitted to them.

Instead, Spanish residents paying tax in Spain can request a *certificado de residencia fiscal NEN – España Convenio,* from their local tax office in Spain. This certificate should then be sent to HMRC Personal Tax International (**see Appendix 1**).

The Spanish authorities will generally want the tax due on the UK pension to actually be paid for the first time (in May/June of the following year) before they will issue a *certificado de residencia.* This might not be until almost 18 months after you have taken up Spanish residence. If this is the case, you will ultimately be repaid any PAYE tax deducted at source in the UK while you were actually resident in Spain; however, you should not then claim double tax relief on your Spanish tax return for the tax paid on the income in the UK.

Spanish taxation: UK state retirement, personal and occupational pension income

If you are resident in Spain, your UK state retirement, personal and occupational pension income will be taxable **only** in Spain, and not in the UK, under the terms of the UK/Spain double tax treaty.

An allowance (between €2,652 and €4,080 in 2013) is available when calculating the taxable income, and tax is payable at the Spanish income tax scale rates.

However, the UK will continue to deduct tax at source from such income (except for the UK state retirement pension, which is always paid gross) until they are satisfied that you are both resident in, and paying tax on this income in Spain.

Although the state pension is paid to you gross (ie, with no deduction of tax at source) sometimes it can be taxed indirectly if you have another source of income subject to PAYE (check your tax code number carefully).

Spanish taxation: UK Government service pensions

If a pension arises from UK Government service, UK tax will **always** be payable even if you are tax resident in Spain, unless there has been a transfer out before the pension commences (and usually before age 59), or you are a Spanish national.

UK Government pensions are not taxed in Spain nor do they count in assessing the level of your income when calculating the rate of tax payable.

Government service pensions are civil service or local authority pensions. They may also be teachers, police and fire brigade pensions. However, this does not always apply, and therefore you should take advice on the matter. Please note that an NHS pension does not count as government service.

The HMRC website publishes a list of pension types, identifying which are Government or non-Government for the purposes of the pensions articles of the double tax agreements in various countries.

Spanish taxation: personal pension income from a source outside the EU

It is possible that income from non-EU pension funds can be taxed as a purchased annuity in Spain. Annuities are taxed differently depending on whether the annuity is 'whole of life' or temporary. Within each category, the annuity can be immediate or deferred. While the income tax treatment of annuities is very beneficial in Spain, this does expose the capitalised value of the annuity to wealth tax. Even so, the overall tax treatment may still be more beneficial than the occupational pension treatment. As you cannot change the position once applied, it is important to review the position carefully before deciding if this is the correct route for you.

Immediate whole of life annuities *(rentas vitalicias inmediatas)*

The taxable income element of the annuity is determined by applying a fixed percentage to the amount received, depending on the age of the beneficiary at the time the annuity vests. Only the fixed percentage of the income is taxable in Spain with the remainder being tax-free.

Deferred whole of life annuities *(rentas vitalicias diferidas)*

The same percentages apply to the income from a deferred annuity but, in addition, tax is also due on the accrued income up to the date the annuity vests (ie, the income accrued during the period of deferral).

The accrued income is calculated as the difference between the present financial-actuarial value of the annuity and the total amount of premiums paid. However, the beneficiary of the annuity is not taxed on this income at the time the annuity vests, as this could present them with a cashflow problem by having to pay tax on income which they have not yet received.

Instead, in the case of a whole of life deferred annuity, the taxable income is regarded as accruing evenly throughout the first 10 years of receipt of the annuity.

Immediate temporary annuities *(rentas temporales inmediatas)*

Depending on the term of the temporary annuity, a fixed percentage of the income will be tax-free dependent on the terms of the annuity.

Deferred temporary annuities *(rentas temporales diferidas)*

The same percentages apply for a deferred immediate annuity, but, in addition, tax is also due on the accrued income up to the date the annuity vests (ie, the income accrued during the period of deferral).

Spanish taxation of the lump sum

This is taxable in Spain if received while Spanish tax-resident (unless it arises from a UK Government service pension). So if you wish to avoid the Spanish tax, you should either receive it before you become a tax resident of Spain or, if you believe you will qualify for the much reduced annuity tax system, simply do not take the lump sum as you will be better off with a bigger pension.

In making this decision, you should bear in mind how many years you must take the additional pension for in order to 'recoup' the amount of the lump sum; whether the lump sum, if invested, could bring you a larger return, or whether your life expectancy may be a factor, if you have been ill, etc.

22.8 UK state pension and UK National Insurance Contributions

Even if you have earnings in Spain, they will not enhance your entitlement to UK state pension or other benefits, and it may be worthwhile for you to consider paying voluntary Class 2 or Class 3 NI contributions in the UK while you are abroad, depending upon how long it is before you reach State retirement age. Neither of these gives cover for health care abroad in any circumstances.

Class 2 contributions protect entitlement to retirement pension, bereavement benefits and invalidity benefit (and maternity allowance), whereas Class 3 contributions only protect entitlement to retirement pension and bereavement benefit. You can only pay Class 2 contributions if you are employed or self-employed abroad. In order to decide whether it is worth your while making voluntary contributions, you should contact the UK National Insurance Contributions Office.

Class 2 Voluntary NIC contributions for those employed or self-employed abroad are at the same rate as the normal Class 2 NIC contributions for those self-employed in the UK, but Class 3 voluntary NIC contributions are at a much higher rate.

Qualifying conditions for full UK state retirement pension

The number of qualifying years for a full State retirement pension is 30 years for both men and women reaching state pension age on or after 6 April 2010, and there is no minimum number of qualifying years before any pension at all can be paid. This is likely to increase over the coming years as the age at which an individual can claim the UK state retirement pension increases.

It is now even more likely that an individual could already have (or could expect to have in future) sufficient contribution years to be eligible for a full State pension without paying any voluntary NI contributions.

The State pension age for both men and women is to increase from 65 to 68 between 2024 and 2046, with each change phased in over two consecutive years in each decade. The first increase, from 65 to 66, will be phased in between April 2024 and April 2026; the second, from 66 to 67, will be phased in between April 2034 and April 2036; and the third, from 67 to 68, between April 2044 and April 2046.

UK state pension deferral

Since 6 April 2005 you have been able to defer taking State pension, or choose to stop claiming it after you have already claimed it. This allows you to build up extra income or a taxable lump-sum payment.

To get extra State pension, you have to put off claiming for at least five weeks. You build up extra pension at 1% of your normal weekly rate for every five weeks you put off claiming (equivalent to about 10.4% extra for every full year you put off claiming). To get a lump-sum payment you have to put off claiming for a continuous period of at least 12 months, and in this case you will get interest at 2% above the Bank of England base rate.

You can find out how much you could get by putting off claiming your own State pension by phoning the Pension Forecasting Team (**see Appendix 2**).

However, deferring the State pension will mean that you will not be entitled to Form S1, which in turn will entitle you to Spanish State health care. This is because only those actually in receipt of their State pension are entitled to the form. Therefore, if you do defer your UK State pension, you need to be prepared to pay for private health care in Spain. It is sensible to compare the extra amount of pension received with the costs of private health care.

22.9 QROPS

Introduction

One of the big changes brought in by the Pensions Simplification Act was the ability to transfer your pension outside of the UK into a Qualifying Recognised Overseas Pension Scheme (QROPS).

Backed by legislation, HMRC has put in place a pre-approved system whereby UK pension rights can be transferred into a QROPS at the member's request. To obtain QROPS status, a QROPS provider must meet a number of the HMRC's reporting requirements for a minimum of 10 complete tax years after the transfer to the QROPS, and the QROPS remains subject to UK charges for at least five years after the member has left the UK.

In order to transfer UK pension rights to a QROPS, the member must have left or intend to leave the UK and become non-UK resident for tax purposes. Provided that a member meets this test UK pension rights can automatically be transferred outside the UK in the same way as pensions can be transferred between approved providers within the UK. Correspondingly, when benefits are transferred to a QROPS this is done without any UK tax charge (unless there is a liability to an excess fund charge, though this is reduced from 55% maximum to 25%; this is described below).

UK pension rights may be transferred outside the UK either before the member commences benefit or potentially once benefits have come into payment. This includes most types of pension including income drawdown currently in payment and protected rights, which are pensions which have accrued as a result of UK national insurance rebates. However, it is not possible to transfer into a QROPS an entitlement to the basic UK State pension or after an annuity has been purchased or from most final salary schemes where payments have commenced.

The unauthorised payment charge

The UK tax authorities can impose a tax charge, called the unauthorised payment charge (also known as 'member payment charge'). The maximum rate is 82% and it applies to any 'unauthorised' payments made by the pension trustees. Such unauthorised payments include payments in excess of the normal approved tax-free lump sum, or a transfer into a more flexible trust. The unauthorised payment charge applies to the overseas QROPS as well as to UK-based pension providers. In other words, it seeks to maintain the restrictive manner in which benefits can be drawn from pensions.

An unauthorised payment would not be made by any UK pension provider because it would be in contravention of the trust deed, which in the UK is tightly drawn in order to obtain HMRC approval in the first place. In any case, any UK pension provider who permits an unauthorised payment is likely to lose their authorisation to manage pension funds, and thus be put out of business. A transfer to a QROPS, however, is not an unauthorised payment.

In order for an overseas QROPS provider to be acceptable to HMRC, they must agree to provide information immediately to the UK tax authorities should there be an event following the transfer that leads to an unauthorised payment, in which case HMRC will assess the member (you) and they can now enforce tax collection across borders.

There is, however, an important exception under Clause 2 Schedule 34 of the Finance Act 2004. This states that where an unauthorised payment is made to an individual who is not a UK tax resident, and has not been so for the immediate previous 10 UK tax years, there is no unauthorised payment charge reporting required, nor is there an unauthorised payment due. Importantly, this includes a transfer into a flexible protected trust where you not only have greater flexibility in the manner that you take your benefits, but also means that the pension fund is entirely out of the UK HMRC's reporting and tax scope.

Benefits of QROPS

As a Spanish resident the benefits following transfer to QROPS are considerable:

- No UK tax is due on any benefits received. Importantly, there is no UK PAYE applicable, whereas if you do not move the fund, UK tax will remain payable on any benefit subject to the terms of any appropriate double tax treaty.
- No tax is payable on income or gains within the fund, other than the inevitable withholding taxes applicable on certain investments.
- You will not be required to buy an annuity and strategies can be put in place so that you are not necessarily restricted to the Government Actuary's Department (GAD) rules for income drawdown or to have a UK alternatively secured pension (ASP) with the resultant tax charge on death once you have been non-UK resident for more than five complete consecutive tax years. However, should you at any stage wish to consider buying an annuity, (ie, following the death of either one of you), that option remains open to you.
- There is no upper age limit at which you have to take benefit.
- The benefit of the fund can pass to your heirs, and it will be inheritance tax protected.
- Your income can be in Euros or any other currency, and is not limited to UK Sterling.
- The export of the pension fund does not mean that you have to be employed overseas, nor indeed do you have to retire or form a company.
- You can have a flexible investment plan across a wide range of funds with a risk profile that matches your attitude to risk. This should be regularly reviewed by your financial advisers so as to ensure that it continues to meet your risk profile and changing circumstances. Importantly, you can take income by way of withdrawals (which need not be in equal or regular amounts) in a more tax-efficient manner than is currently available to you.

- Should you ultimately return to the UK, only 90% of the pension income would be taxable in your hands (as this is a foreign source pension).

Regulation

Should you decide to start looking at transferring your pension, with a view to maximising both the amount of income that you might want to take, and the degree of flexibility available to you, you should proceed with caution.

Pension transfers are heavily regulated by the Financial Conduct Authority (FCA) in the UK. As you may become a Spanish resident then careful consideration should be given to only taking pension transfer advice from advisers who are both regulated in the UK and in Spain and are specifically authorised to give advice on transferring UK pensions. In the event that an adviser is not authorised on this basis, then you cannot be sure about the competency of the advice given, and you may not be protected by the professional indemnity cover of the adviser.

22.10 QNUPS (Qualifying Non-UK Pension Scheme)

These schemes allow UK residents to transfer assets into a pension scheme without any relevant earnings. They are treated as pension schemes for UK purposes, and there are rules about the amounts that may be withdrawn, as for all pension schemes. They have certain advantages:

- You can draw the 25% pension commencement lump sum tax-free as a UK resident (although this is taxable if you are resident in Spain at the time).
- The pension fund immediately falls outside the UK inheritance tax net.

These are genuine pension schemes, designed to provide you with a pension income, and if there is any suspicion that the scheme has been entered into with the sole intention of avoiding UK inheritance tax, it will fail. This could happen where the individual already has sufficient pension income, or is over around 75 or has a low life expectancy.

There are disadvantages of putting non-cash assets into such a scheme, as the assets are effectively 'sold' to the scheme in exchange for the future pension income (which is why they are not subject to UK inheritance tax). Therefore, if non-cash assets are transferred to the scheme, capital gains tax may be due in your country of residence.

There may be scope for receipts from such a scheme to be treated as an annuity in Spain, but this does expose the capitalised value of the scheme to wealth tax, so advice should be taken.

"The trouble with retirement is you never get a day off!"

ABE LEMONS

QROPS

QROPS are often focused on in the press but are you sure they are right for you?

We evaluate what is right for you, if that is a QROPS then that is fine, but QROPS are an optional solution and should not be the starting point.

We need to talk.
www.blevinsfranks.com

APPENDIX 1

Glossary of Spanish Terms

Spanish	**English**
A	
Abogado/a	Lawyer
Abusos Urbanisticos NO	Pressure group against land seizures
Acta	Minutes
Actos Juridicos Documentados	Stamp duty
Administración	Local government administration office
Agencia Tributaria	Spanish tax office
Agente de la Propiedad Inmobiliaria (API)	Estate agent
Albacea	Executor
Alcalde	Mayor
Asesor fiscal	Tax adviser
Audiencia nacional	National court
Audiencia provincial	Provincial court

Audiencia regional	Regional court
Autónomo	Independent
Ayuntamiento	Town hall

B

Baja	(Off the road) car document
Base imponible	Taxable amount/base
Base liquidable	Net taxable base
Basura	Refuse tax
Boletín Oficial del Estado	Official State Journal
Bonificación	Bonus

C

Cadastral	Register
Caja postal de ahorros	Post office (savings bank)
Caja de ahorros	Bank
Carnet de conducir	Driver's licence
Cartilla de ahorros	Savings bank book
Catastro/al	Office for property valuation
Cédula de habitabilidad	Habitation certificate (to certify that a property is habitable)
Cédula de ocupación	Occupation certificate
Centro de Estudios Financieros	Financial training centre
Centro de reconocimiento médico	Medical examination centre
Correo certificado	Registered mail

Certificado de antecedentes penales	Certificate of criminal record
Certificado de equivalencia	Certificate of equivalence
Certificado de fin de obra	Completion certificate
Certificado de identificación fiscal (CIF)	ID tax number
Certificado de residencia fiscal NEN	Certificate of residence tax number
Certificado internacional de seguro de automóvil	International car insurance certificate
Certificado de ultimate voluntades	Certificate of ultimate Wills
Comisaría	Police station
Comunidad Autónoma	Autonomous community (AC)
Comunidad de Propietarios	Land/home owners' association (community of owners)
Comunidad Valenciana	Valencian community
Congreso de los Diputados	Congress of Deputies (Parliament)
Consejo de estado	Council of state
Contrato privado	Private contract
Contribución urbana	Town rates
Coordinación Provincial de Donantes de Transplantes	Organ donor organisation (in Málaga Province)
Cuenta ajena	Employed
Cuenta extranjera	Foreign account
Cuota de abonano	Standard charge for telephone line
Cuota de participación	Community expenses

D

Declaración de obra antigua	Declaration of previous work
Declaración de obra nueva	Declaration and registration of a new build
Declaración de siniestro de automóvil	Accident report form
Declaración ordinaria	Standard tax return
Declaración simplificada	Simple tax return
Defensor del pueblo	Ombudsman
Delegación	Local head tax office
Delegado de trabajo	Provincial director of the labour ministry
Demanda	Civil action
Denuncia	Formal complaint
Departamento de extranjeros	Foreigner's department
Departamento de urbanismo	Urban department
Despido improcedente	Unjust dismissal
Diferida	Deferred
Dirección general de seguros (DGS)	Spanish insurance regulator
Dirección general de tráfico	Traffic department
Dirección general de transacciones exteriores	Foreign exchange authority
Director de pompas funerales	Funeral director
Domiciliación de pagos	Standing order

E

Empresario individual	Sole trader
Escritura	Title deeds
Escritura de cancelación	Cancellation in writing
Escritura de compraventa	Contract of sale
Escritura de propiedad	Title deed
Escritura pública	Formal publicly registered title deed
Estado de cuenta	Bank statement
Estatal	State
Estatuto de los Trabajadores	Workers' statute
Estatutos	Rules

F

Facturación	Period of billing
Finca	Old farmhouse/property
Fiscalidad internacional	International tax
Franquicia	Franchise

G

Gestor	Business agent
Gestor Intermediario en Promociónes de Edificiones (GIPE)	Estate agents' licensing body
Gestoria	Agency undertaking business with government departments, insurance, etc
Guardia Civil	Civil guard

H

Hacienda	Spanish tax office
Hecho imponible	Taxable event
Herederos forzosos	Obligatory heirs
Hipoteca	Mortgage
Hogar de pensionistas	Spanish pensioners' club
Hoja de reclamación	Complaint sheet

I

Impuesto General Inderecto de Canarias (IGC)	Canarian indirect general tax
Instituciones de Inversión Colectiva (IIC)	Spanish approved public access collective investment vehicles that invest in unquoted investment funds
Ilimitado	Unlimited (amount)
Imponible	Taxable base
Impuesto especial Sobre Determinados Medios de Transporte (ISDMT)	Registration tax that cover cars, boats and aircraft
Impuesto Sobre actividades económicas	Business tax
Impuesto Sobre Actos Juridicos Documentados (AJD)	Stamp duty
Impuesto Sobre Bienes inmuebles (IBI)	Property tax
Impuesto sobre el incremento de valor de los terrenos de naturaleza urbana (plusvalía)	Urban land tax/capital gain

Impuesto sobre el patrimonio	Wealth tax
Impuesto sobre el Valor Añadido (IVA)	Valued added tax
Impuesto sobre la circulación de vehículos	Local car tax
Impuesto sobre la renta de las personas físicas	Personal income tax
Impuesto sobre sociedades	Corporate tax
Impuesto sobre Sucesiones y Donaciones (ISD)	Inheritance and gift tax
Impuesto sobre Transmisiones Patrimoniales (ITP)	Spanish property transfer tax
Incendio y robo	Fire and theft
Indemnizaciones de despido	Termination or serverance payments
Índice de Referencia de Préstamos Hipotecarios (IRPHCE)	Index of reference of hypothecating loans
Información impresa	Printed information
Inmediata	Immediate
Inspección técnica de vehículos	Roadworthiness test
Instituto de Migraciones y Servicios Sociales (IMSERSO)	Social security authorities
Instituto Nacional de Empleo (INEM)	National employment office
Instituto Nacional de Seguridad Social (INSS)	Social security office
Inventarios y cuentas anuales	Annual accounts

J

Jefatura provincial de tráfico	Provincial traffic department
Juez de primera instancia e instruccíon	Coroner
Junta de fiscales de sala	Meeting of public prosecutors
Juzgado de distrito	District court
Juzgado de paz	Justice of the peace
Juzgado de primera instancia	First appeal courts
Juzgados centrales de instrucción	Central examining magistrate

L

Las parejas estables	Unmarried couples within a stable relationship
Ley	Law
Licencia de Primera Ocupación	Licence of first occupation
Licencia fiscal	Financial licence (registration with the Spanish Inland Revenue formerly required for any commercial activity)
Liquidable	Net base

M

Magistratura de Trabajo	Industrial tribunal
Mensualidades	Monthly pay
Minimo	Basic allowance
Minimo personal y familiar	Personal tax free allowance
Ministerio de Economia y Hacienda	Treasury

Módulos	Modules
Multa	Penalty
Municipio	Town council

N

Nota simple	Simple report
Notario	Notary
Nuda propiedad	Free holder interest/reversionary interest
Número de Identificación (NI)	Fiscal tax identify number
Número de Identificación de Extranjeros (NIE)	Tax identity number for foreigners
Número de Identificación Fiscal (NIF)	Tax identity number

O

(Oficina de) correos	Post office
Oficina de extranjeros	Foreigners' office
Oficina municipal de información al consumidor	Consumer information/ complaints office

P

País Vasco	Basque country
Papel del estado	Government bonds
Parador	State-run hotel
Pareja de hecho	A *de facto* couple

Pareja estable	Unmarried couple in a stable relationship
Partido Popular (PP)	Party of the People (equivalent to the UK Conservative Party)
Partido Socialista Obrero Español (PSOE)	Socialist/Workers' Party
Patrimonio	Wealth
Pecunia	Cash
Permanecer	To stay or to remain
Permiso de circulación	Document for driving on public road
Permiso de residencia	Residence permit
Plan General de Ordenacion Urbanistica (PGOU)	Equivalent to the UK town urban plan
Plaza major	Main square
Plusvalía	Land value tax (lit: capital gain)
Poderes	Power of attorney
Policia nacional/municipal	National/local police
Por temporada	Short term holiday lets
Presidente	Prime Minister or chairman

R

Real Automóvil Club de España	Spanish equivalent to the RAC
Regimen Economico y Fiscal (REF)	Special tax regime
Regímenes especiales	Special regime for social security contributions
Registro de la Propiedad	The land registry

Registro de Parejas Estables	Registry for unmarried couples in stable relationships
Registro Mercantil	Companies' registry
Reglamento de Ordenancion y Supervisión de los Seguros Privados (ROSSP)	Private insurance regulation and supervision code
Renta del ahorro	Savings income
Renta general	General income
Renta vitalicia	Whole of life annuity
Rentas vitalicias inmediatas	Immediate whole of life annuity
Rentas vitalicias diferidas	Deferred whole of life annuity
Representante fiscal	Taxpayer's agent
Residencia	Residence
ROEZEC	Official Register of Entities of the Canary Islands

S

Segunda vivienda	Second house
Seguridad social	Social security
Seguro de responsibilidad social obligatorioa	Third party insurance
Seguro occupante	Driver and passenger insurance
Senado	Senate
Sistema bonus-malus	No claims bonus
Sociedad anónima	Limited liability company
Sociedad colectiva	Partnership
Sociedad limitada	Private limited company

Solicitud de carnet de permiso de conducir	Application for driving licence
Solicitud de datos de registro de vehículos	Request for vehicle registration details
Solicitud de franquicia	Franchise application
Sucursal	Branch office, subsidiary
Suma	Local tax office

T

Tarjeta comunitaria	Community card
Tarjeta de inspección técnica de vehículos	Certificate of road worthiness
Tasas	Rates
Telefónica	Spanish national telephone service
Tenientes de alcalde	Deputy mayors
Terreno rústico	Rural land
Terreno urbanizable	Building/development land
Tesoreria General de la Seguridad Social (TGSS)	The general treasury of social security
Testaferros	Nomineeship
Testamento abierto	Open Will
Todo riesgo	Fully comprehensive car insurance
Todos los gastos	All the expenses
Total a pagar	Total to be paid
Transferencia	Transfer
Tribunal Central de Trabajo	Central Industrial Court

Tribunal Constitucional	Constitutional Court
Tribunal Económico	Central Economic Administrative Court
Tribunal Supremo	Supreme Court

U

Union de hecho	Union in fact (married couples)
Union de pareja	Unmarried couple
Urbano	Urban
Urgente	Urgent
Usufructo/s	Life interest (usufruct)
Usufructorio	Life interest holder

V

Valor catastral	Property value
Vehículos	Vehicle
Vivienda habitual	Habitual housing
Vivienda secundaria	Secondary residence
Viviendas de Protección Oficial (VPO)	Government assistance built property (council housing)

Z

Zona Especial Canaria	Special zone, Canary Islands (for tax purpose)

APPENDIX 2

Useful Names and Addresses

Blevins Franks Financial Management Limited
28 St James's Square
London SW1Y 4JH
United Kingdom
Tel: +44 (0)20 7389 5220
Fax: +44 (0)20 7389 5230
Email: enquiries@blevinsfranks.com

Blevins Franks Financial Management Limited
Gasan Centre, Triq il-Merghat,
Mriehel BKR3000
Malta
Tel: +356 2327 4000
Fax: +356 2134 7348
Email: enquiries@blevinsfranks.com

COMMERCE

Dirección General de Seguros y Fondos de Pensiones (DGSFP)
Paseo de la Castellana, 44
28046 Madrid
Tel: +34 902 171 111
Fax: +34 913 397 113
www.dgsfp.mineco.es

Financial Conduct Authority (FCA)
25 The North Colonnade
Canary Wharf
London E14 5HS
Tel: +44 (0)20 7066 1000
www.fca.gov.uk

Spanish Chamber of Commerce in Great Britain
126 Wigmore Street
London W1U 3RZ
Tel: +44 (0)20 7009 9070
Fax: +44 (0)20 7009 9088

COMMERCE

Email: info@
spanishchamber.co.uk
www.spanishchamber.co.uk

TTT Moneycorp Ltd
100 Brompton Road
Knightsbridge
London SW3 1ER
Tel: +44 (0)20 7589 3000
Email: enquiries@
moneycorp.com
www.moneycorp.com

GOVERNMENT DEPARTMENTS

British Consulate – Alicante
Edificio ESPACIO
Rambla Nendez Nunez 28-32
6a Planta
03002 Alicante
Tel: +34 902 109 356/
+34 917 146 300
Fax: +34 965 140 528
Email: info.consulate@fco.gov.uk

British Consulate General – Barcelona
Edificio Torre de Barcelona
Avenida Diagonal 477 13°
08036 Barcelona
Tel: +34 902 109 356/
+34 917 146 300
Fax: +34 933 666 221
Email: info.consulate@fco.gov.uk

British Consulate – Bilbao
Alameda de Urquijo 2-8
48008 Bilbao
Tel: +34 944 157 722
Fax: +34 944 167 632
Email: bilbaoconsulate@
fco.gov.uk

British Council – Bilbao
Avenida Lehendakari Aguirre 29,
2°
48014 Bilbao
Tel: +34 944 763 650
Email: bilbao@ britishcouncil.es

British Vice Consulate – Ibiza
Avenida Isidoro Macabich 45
Apartado de Correos 307
07800 Ibiza
Tel: +34 900 109 356/
+34 317 146 300
Fax: +34 971 301 972
Email: info.consulate@fco.gov.uk

GOVERNMENT DEPARTMENTS

British Consulate General – Madrid
Torre Espacio
Paseo de la Castellana 2590
28046 Madrid
Tel: +34 902 109 356/
+34 917 146 300
Fax: +34 917 146 403
Email: info.consulate@fco.gov.uk

British Council – Madrid
Paseo del General
Martínez Campos 31
28010 Madrid
Tel: +34 913 373 577
Fax: +34 913 373 573
www.britishcouncil.org

British Embassy – Madrid
Torre Espacio
Paseo de la Castellana 2590
28046 Madrid
Tel: +34 902 109 356/
+34 917 146 300
Fax: +34 917 146 301
Email: info.consultae@fco.gov.uk

British Consulate – Málaga
Edificio Eurocom, Bloque Sur
C/Mauricio Moro Pareto, 2
29006 Málaga
Tel: +34 902 109 356/
+34 917 146 300
Fax: +34 928 267 774
Email: info.consulate@fco.gov.uk

British Consulate – Las Palmas de Gran Canaria (Canary Islands)
Calle Luis Morote 6, 3°
35007 Las Palmas de
Gran Canaria
Tel: +34 902 109 356/
+34 917 146 300
Fax: +34 928 267 774
Email: info.consulate@fco.gov.uk

British Consulate – Palma de Mallorca
Convent dels Caputxins No. 4
Edificio Orisba B 4th D
07002 Palma de Mallorca
Tel: +34 902 109 356/
+34 917 146 300
Fax: +34 971 717 520
Email: info.consulate@fco.gov.uk

British Consulate – Santa Cruz de Tenerife
Plaza Weyler 8, 1°
38003 Santa Cruz de Tenerife
Tel: +34 902 109 356/
+34 917 146 300
Fax: +34 922 289 903
Email: info.consulate@fco.gov.uk

British Council – Valencia
Avinguda de Catalunya 9
46020 Valencia
Tel: +34 963 392 980
Email: valencia@
britishcouncil.es
www.britishcouncil.org

GOVERNMENT DEPARTMENTS

Enquiries about your **UK NI contributions** should be made to:
HM Revenue & Customs
UK National Insurance Contributions Office
International Services
Benton Park View
Longbenton
Newcastle Upon Tyne
NE98 1ZZ
Tel: +44 (0)19 1203 7010

HMRC Residency
UK Tel: 0300 200 3300
from outside the UK:
+44 (0)13 5535 9022
www.hmrc.gov.uk

To Apply for Form S1:
Department for Work & Pensions (DWP)
Overseas Medical Benefits Section
The Pension Service
International Pension Centre
Tyneview Park
Newcastle Upon Tyne
NE98 1BA
Tel: +44 (0)19 1218 1999
Fax: +44 (0)19 1218 7381
www.thepensionservice.gov.uk

If you need help about **benefits** write to:
Overseas Benefits Directorate
Department for Work & Pensions (DWP)
The Pension Service
International Pension Centre
Tyneview Park
Newcastle Upon Tyne
NE98 1BA

Instituto Nacional de la Seguridad Social
Subdirección General de Relaciones Internacionales (DSS)
Padre Damian 4
28036 Madrid
Tel: +34 915 688 300
Fax: +34 915 640 684
www.seg-social.es

If you live (or have lived) in Northern Ireland write to:
Department for Social Development
Lighthouse Building
1 Cromac Place
Gasworks Business Park
Ormeau Road
Belfast BT7 2JB
Tel: +44 (0)28 9082 9000

Spanish Consulate General – London
20 Draycott Place
London SW3 2RZ
Tel: +44 (0)20 7589 8989
Fax: +44 (0)20 7581 7888
www.conspalon.org

GOVERNMENT DEPARTMENTS

Spanish Consulate General – Scotland
63 North Castle Street
Edinburgh EH2 3LJ
Tel: +44 (0)13 1220 1843
Fax: +44 (0)13 1226 4568
Email: cgspedimburgr@
mail.mae.es

Spanish Embassy – UK
39 Chesham Place
London SW1X 8SB
Tel: +44 (0)20 7235 5555
Fax: +44 (0)20 7259 5392
Email: admin@
embassyhomepage.com
spain.embassyhomepage.com

UK in Spain
www.ukinspain.com/english

HEALTH CARE SERVICES

AXA PPP Healthcare
Philips House
Crescent Road
Tunbridge Wells
Kent TN1 2PL
Tel: +44 (0)87 0608 0850
www.axappphealthcare.co.uk

BUPA International
Victory House
Trafalgar Place
Brighton BN1 4FY
Tel: +44 (0)12 7332 3563
Fax: +44 (0)12 7382 0517
Email: info@bupa-intl.com
www.bupa-intl.com

PENSIONS

The Pension Service
Advice on transfer of pensions overseas. Advice on medical cover and making claims overseas.

International Pension Centre
Tyneview Park
Newcastle Upon Tyne
NE98 1BA
Tel: +44 (0)19 1218 7777
Fax: +44 (0)19 1218 7381
www.thepensionservice.gov.uk
(Application form to use: BR19)

PET TRAVEL AND INSURANCE

PetPlan
Great West House (GW2)
Great West Road
Brentford
Middlesex TW8 9DX
Tel: +44 (0)84 5077 1934
www.petplan.co.uk

SCHOOLS

British Education in Spain
NABSS is the National Association of British Schools in Spain and was founded in 1978. Each of its member schools is subject to inspections from the UK every five years. There are over 30 member schools and other British schools in Spain which are not members but are affiliates.

National Association of British Schools in Spain (NABSS)
Comercio 4-Esc. 2-Bajo-A
28007 Madrid
Tel: +34 915 520 516
Fax: +34 915 520 517
www.nabss.org

TOURISM

Spanish Tourist Office
6th Floor
64 North Row
London W1K 7DE
UK Tel: +44 (0)20 7317 2011
www.spain.info
www.tourspain.co.uk

APPENDIX 3

Advice and Guidance

Anyone moving overseas has a lot of financial restructuring to deal with and they often take advice from their UK advisers who are not well placed to deal with such enquiries and cannot maintain the ongoing advice once their clients are abroad.

Equally, local advisers in Spain are unlikely to understand, how and when UK tax can still impact a UK expatriate, nor the implications of the double taxation treaty which governs the interaction between the UK and Spanish taxation regimes.

In addition, pension issues abroad are different to the UK, taxes are often higher and investments that work in the UK are not suitable for retired overseas residents.

It is important to ideally seek advice well in advance of going overseas. Having someone on your side that not only speaks the language, but understands the local laws and has the knowledge to deal with the bureaucracy will make the whole transition easier.

For UK nationals who have already moved to Spain, there are a number of important tax and wealth management issues to consider. It is again important to be advised by someone who understands the recent changes to Spanish taxation, UK taxation and the interaction of the two regimes.

At Blevins Franks we pride ourselves in providing high quality advice to UK nationals living abroad, or planning to do so.

The list outlined below shows some of the areas we focus on but to fully explore your specific situation we would recommend arranging an appointment with one of our Partners or Private Client Managers.

- Tax residence.
- Income tax planning.
- Treatment of pensions and annuities overseas.
- UK and overseas inheritance tax planning.
- Succession tax and 'forced heirship' rules.
- Domicile.
- Offshore trusts.
- Capital gains tax – UK and overseas.
- Property ownership structures.
- Taxation of second homes.
- Tax efficient investments.
- Social security as applied in your country of interest.
- Health care arrangements overseas.
- Corporate structures.
- Rental income.
- Double tax treaties.
- Legal tax avoidance both in the UK and overseas.
- The effective and appropriate use of tax havens

For advice and guidance tailored to you and your family's specific situation and needs, please contact us using the following details.

Southern Spain

+34 900 460 174	Nerja, Málaga, Mijas, Marbella, Nueva Andalucía
+34 952 809 212	San Pedro to Cádiz
+34 952 799 752	San Pedro to Cádiz

Eastern Spain

+34 965 793 561	Valencia to Moraira
+34 965 705 502	Alicante (South) to Almeria
+34 933 443 316	Costa Brava

Spanish Islands

+34 971 719 181	Balearics (Mallorca South)
+34 971 867 781	Balearics (Mallorca North)
+34 922 716 079	Canary Islands

Other Areas of Spain

+44 (0)20 7389 5220	London

Index

FIGURES REFER TO CHAPTER SECTIONS